The Australian Guerilla 4.

TRAPPING THE JAP

Ion Idriess

ETT IMPRINT
Exile Bay

I0066820

This edition published by ETT Imprint, Exile Bay 2020

Also by Ion Idriess
Shoot to Kill
Sniping
Guerrilla Tactics
Lurking Death
The Scout

First published 1942 by Angus & Robertson
Facsimile edition published by Idriess Enterprises 1999
Electronic edition published by ETT Imprint 2020

ISBN   978-1-922473-13-4   (pback)
ISBN   978-1-922473-14-1   (ebook)

ETT IMPRINT
PO Box R1906
Royal Exchange NSW 1225
Australia

Designed by Tom Thompson

CONTENTS

DARWIN HEAVILY BOMBED

ATTACKS BY 93 PLANES: 4 SHOT DOWN

DAMAGE "CONSIDERABLE": CASUALTIES UNKNOWN

DARWIN WAS HEAVILY BOMBED BY 93 JAPANESE PLANES IN TWO RAIDS YESTERDAY.

Mr. Curtin, Prime Minister, announced last night that the first attack was made by 72 twin-engined bombers, accompanied by fighters. The second was by 21 twin-engined bombers.

On February 18 1942, 242 Japanese aeroplanes bombed Darwin, inflicting many casualties and damage, prompting Idriess to write the Guerrilla Series. Over the next year a further 110 Japanese raids on Australia and its shipping galvanised the country into preparing for imminent invasion.

I
Bog that Mechanised Spearhead - Smash it!

IN *Guerrilla Tactics* we realized we were up against a ruthless, efficient, brave, well-equipped, and very numerous foe. We decided that the way to beat him was to use our brains as well as ever growing armaments. We thought out a few ways of beating him, calling to our aid cunning, daring, and the forces of nature to help us against his armaments and numbers. By now you should have thought out other tricks, other methods of surprise and daring attack, for if you guerrillas have discussed that little book it will have sharpened your wits. One idea leads to another, one subterfuge to another, one good plan to a plan still better.

We will now get to work and plan out a few more, for the time is drawing very close; the enemy is very near; he has struck at our islands and overwhelmed some; our Allied airmen and our navies have come to death-grips with him. Our boys are not numerous away out there; they may not be able to hold him off our shores for long.

In his too numerous victories, so far, the enemy has rushed to his exceedingly well-planned attack after careful aerial reconnaissance followed by preliminary bombing. Partcularly against aerodromes and grounded aircraft, for he well knows the danger of land-based aircraft against him. And he greatly desires those air bases for himself. He plans his attacks to capture them.

After bombing dromes he turns his planes against troop concentrations, artillery, tanks, and strong-points which would be strategically placed to oppose him. His object has been to choose the most advantageous landing-place; destroy as many allied planes as possible; disorganize. the defences. Then in he rushed, in numerous armour-plated barges, many of them high powered, each carrying from fifty to two hundred well-armed men. In good offensive and protective formation these sped ashore from the troopships under cover of heavy aerial and naval bombardment. Losses do not count with this enemy so long as he gains his objective.

He attacks in force; tries to push his way through. If he doesn't succeed quickly, he tries infiltration, which we have already discussed. A favourite thrust of his is to rush a spearhead of tanks, armoured cars, and lorry troops through a weak point and on into the country far beyond.

Such a spearhead may come tearing through your locality.

Can you stop it? Or at least do something about it? A few traps have been suggested in *Guerrilla Tactics*. Perhaps the local conditions of your country may not lend themselves to the methods outlined. But the local peculiarities of your district may possibly offer you other opportunities.

Is there any boggy country in your locality?

In various States, in certain localities, there are from small to considerable areas of boggy ground. In some areas these boggy patches last for a considerable time after rains, They have an apparently firm crust, but in places a horse may sink to its saddle-girths. An ordinary car goes down to the axle.

What would happen there to a heavy lorry or an armoured car, let alone a tank?

A country road may run right through a boggy patch. Slew a yard off the made road and down you go. Ask the teamsters, the timber-getters, the prospectors, the station men in your band. There are sure to be a few men among you whose work takes them intimately into the back country.

Do you know any such boggy patch or patches? You should do so by now; you should know everything that is to be known about your locality. Never forget that an intimate knowledge of every position of your district is one of your greatest weapons.

If an armoured enemy spearhead came bowling along such a road, could you lure it off into the bog? Plan it out.

It is easy enough to make a dummy track leading out from the road. Say towards a clump of timber a quarter of a mile away, the enemy come bowling along, keeping a sharp look out. Open country in places, a creek here and there, timber elsewhere. They come bowling around a clump of timber to see a quarter of a mile off the road a group of Australians boiling the billy. One startled glance, a shout, and the Australians run for their horses. Most foolishly these are tethered some distance away. The leading enemy steps on it and swings the wheel to a branch track that leads directly towards the Australians. A couple of hundred yards farther on the leading car suddenly sogs into a bog. The two cars following cannot pull up in time; they swerve. Ah! they are into it too.

And now rifles crack suddenly, apparently from behind fallen logs across to the right - and left too! The crews and the first three armoured cars are doomed. Such is a simple trap that could be laid in practically any area of boggy country.

There are types of country in which that simple plan could be considerably elaborated upon, so as to smash nearly all, if not all, of such a raiding armoured spearhead. It would only be necessary to have 'in the band a few local men who are familiar with the area, who know it in the way that station men would know it. Such men would have been constantly riding and driving over it in the course of their yearly work. They would know the soft patches and the hard patches.

In a boggy area not all of it is bog, as a general rule. Some patches may be awful; an ordinary car would immediately sink to the axles then quickly begin to settle down. In other patches the car would more or less quickly break through the crust, then experience a desperate struggle, and not a little luck, to get out of the fix. Yet other patches would be comparatively hard; some quite hard. It is easy to understand how such country would be considerably more dangerous to heavy, military wheeled vehicles than to ordinary cars. And ordinary cars dread that country. A heavy tank, once in, would never get out.

The plan would be to lay, at a strategical point on the main road, 'Your dummy track leading off the road and following the hard patches. The track, drawing farther off the road, would lead away just skirting the actual boggy patches linking up with all the hard patches. The plan would be to lure every enemy vehicle off the road deeper into the heart of a morass while at the same time they would have no idea they would have no idea they were going into it. For we have boggy areas that are very deceptive; apparently the side of the track and the country all around looks quite firm, no sign of a bog whatever.

The party of Australians who would clear away in alarm would be a fairly large party of men. When they did reach their horses they would mount and be away at full gallop along the track. They would have to arrange to guard against machine-gun fire from moving tanks or armoured cars. With the targets moving also this would not be nearly so dangerous as otherwise. If there was sufficient cover the men could work it by placing their horses in such positions that all mounted in scattered groups yet at the same time. With individuals widely spaced along the track, and each group wider spaced, flying at the gallop along a track that twisted and wound in and out, and with the pursuing cars having to twist and turn, all might get away without casualties. It would depend on how it was timed for the natural cover, the shape, and the distance along the track. The speed of the cars and horsemen must be taken into consideration. Any man whose horse went down would be safe for he could dive for shelter to either side of the track. The cars would keep on after the main body; if any car slewed off it would be into bog anyway.

The dummy track would lead into the most treacherous portion of the area. Suddenly the galloping men, as if aware they were almost caught, would branch off to right and left. The track would lead straight on. Judging by the country just there, it would immediately appear to the officer in the leading car that, if he kept straight on to a patch of open ground ahead, his cars would be able to wheel to right and left and, with their superior speed, encircle the horsemen in a matter of minutes.

If the trap were well laid, almost certainly that is the

way the officer in charge would think, and shout or wave his orders. Remember, some of these boggy areas are in open or fairly open country, except for occasional patches of timber. The enemy could see all around and satisfy themselves that they were not being led into a trap.

Some cars would almost certainly race straight ahead; others to left and right where the horsemen were vanishing. All would plunge into a morass.

It is quite possible that none would get back. The speed of all three pursuing sections would carry them on. When the first of each bogged those following would be into it, too, before they could pull up. Any following behind would swerve and think they could get around the bogged vehicles. But they too would sink.

Those who raced to right and left would be in the worst fix for they would have no dummy track to guide them back . They would have their own track - if they could get back on to it, and if then it would hold them.

Any traveller who has been caught in such country as that knows what an awful job he has even to get back on to the main road, let alone what his chance would be if lured away like that.

When the trap was sprung the horsemen could wheel back and attack the enemy. They could do it on foot and thus defy the bog. Any enemy who got out of the sheltering vehicles to lend a hand at hauling another out would instantly become a target. The more those vehicles struggled to get out of the bog, the tighter they would stick. They would simply stay there.

If any did pull up in time they would dawdle there while efforts were made to pull the others out. This could not be done for, quite apart from the bog, all enemies who got out of the shelter of their vehicles would be under fire all the time. They could only retaliate from within their tanks, firing at a hidden foe who completely encircled each section. The crews dare not get out and the tanks would have lost their mobility. And those (if any) who had not fallen into the trap, with bog in front and to right and left of them, would not dare leave the dummy track.

While they were in this awful fix 'You could probably get them all, providing you had sufficient men, by blocking the dummy track at some critical place-a spot where to deviate from it either to right or left would mean bog. This could be arranged before the enemy spearhead appeared along the main road. Some of you could be hidden, waiting. Immediately the horsemen dashed off and the cars disappeared in pursuit, hurry to that particular spot and block it.

Any cars or tanks that eventually might come crawling back along the dummy track would find the way blocked. It would then be up to you.

Such a trap would have to be planned carefully, and would depend on a thorough knowledge of the boggy area. One or two men, who were thoroughly familiar with the locality, could carefully show the others the boggy areas and the hard. Every man would have to know the particular part he must play, the party he must work with and follow, and the man or men he must ride behind. The men ahead, who were to deviate to right and left from the dummy track, must lead the horsemen through the bogs; arid they would have to be closely followed, otherwise some of the horses would bog.

Discuss the idea among yourselves; and you'll quickly see how, if ever an enemy comes to pass that way you can trap him.

The dummy track would be easily enough made. When you have carefully selected the route then run a station truck back and along it a few times to make the wheel tracks. Ride your horses back and along it to give it the appearance of a solid, well-used track.

It would be an easy job, except at the very end where the dummy track ends right on an awful bog. You could not run a truck across here of course, nor ride horses over it. But you could take off a wheel and run the wheel tracks across on foot. There are various ways. You must give that last portion of the track the appearance of carrying straight on. It is simple enough to do. Also, you have this fact to help you. The enemy, chasing you all out along the dummy track, would have their attention strictly diverted to you as you separate. They would not give the faintest thought to the continuation of the track ahead except that it offered an obvious opportunity to them of cutting you off. So, although the dummying of the last portion of the track would appear to be most difficult, it actually would be easy.

A different type of country, very gluey, are our blacksoil plains. A nice mess of the roads heavy traffic would make there after rain. Lead a laden enemy convoy into such country in wet weather and the trouble they would have in getting out of it would be just too bad.

In isolated parts of the continent the road here and there crosses the beds of dry creeks and rivers. Occasionally, such a crossing is a trap. Not through water or bog, but because of a type of sand into which the wheels quickly sink. Such a crossing may be from a couple of hundred yards to a quarter of a mile and more across. Probably it is corduroyed. It would be a simple enough matter to remove the corduroy.

There are crossings also, with a steep pull up the opposite bank. Where the road emerges from creek or river the road has been cut out to allow of the ascent. With pick and shovel and elbow grease this tongue could be cut level with the bank, thus leaving the enemy vehicle with a climb up a sheer wall. While they were thinking about it, trying to find a way around or sending back for shovels, a few riflemen could keep them entertained.

Then there is the shallow watercourse where the crossing follows a tortuous rocky bar. On either side mar be a water-hole, or boggy sand, or a form of quicksand. A few nice holes dug into or across the crossing itself would not only block it but would take quite a lot of filling in. This, besides delaying the enemy when he would almost certainly be in a hurry, would give snipers an opportunity to harry his working parties and cause him furious embarrassment.

Another watercourse that I (and you probably) have in mind would afford wonderful cover: islands above and below stream, clumps of scrub, boulders all over the place, heavy cover on the banks. A few handy logs and rocks here and there that the enemy would seize for filling-in purposes would be ideal. You could plant a mine close to each. The enemy would then be scared to touch rock or stick. If you put your hands on a log and it explodes, then your cobbers simply hate the thought of putting their hands on other logs. Yet the enemy must fill in those holes. And he'd have to get out of his tanks and trucks to do it. He would be in plain view as he crawled from his dreadnoughts and all the time he was working. But not one of you would be seen. You could make matters very interesting for him. It would need comparatively few of you to hold up such a convoy indefinitely, if not actually wipe it out.

In strategical places along quite a number of bush roads and at crossings, where a wheeled vehicle is forced to keep to the tracks, a lot of damage could be done with craftily placed mines; either under water or otherwise. It would cause a nasty shock if, for instance, a tank was laboriously climbing up a steep bank when the whole bank blew off in its face. With the mine expertly laid that tank would need quite a lot of digging out - especially if you felled a tree on top of it. You know how those big gums stand up on the bank above the crossings? Well, have one ready, almost cut through.

Or a good mine placed under water in a narrow crossing. Remember, the enemy want speed above all things. Not only would that tank block the way but they would face a big problem to shift it out of the way so that the following vehicles could pass. And it would be just too bad if you could call up planes to come and lay a few eggs in the blockage.

If no planes are available, and the enemy get through, well, you will have delayed them and caused them damage and casualties. They may push you back into the hush. If so, immediately they have passed, blow back to the crossing and wreck again. Do a better job this time. Then, when the next convoy comes to pass over that crossing you will delay it even longer; will cause it more damage, and more casualties.

Oh yes, there's many a trap, and many a repeat trap could be laid in many a way on many an Australian road.

2
Secrets of Tanks

AS aids in helping defeat tanks and mechanized troops we have already called to our help some powerful weapons. Timber, mud, boulders, the old bullock-chain, fencing-wire, etc. They cost nothing; we are not dependent on the nation for their supply. And yet, granted the opportunity, and if they are used in the right Way, these can be formidable weapons. We'll see what we can do with another great natural weapon - fire.

Fire is a terrible weapon. Nothing can stand against it. The aeroplane, the strongest tank, the mightiest battleship, can all be destroyed by fire. But, you must control it, for it is as likely to turn upon friend as upon foe. Australia knows only too well what a terrible thing a bushfire is. It could destroy an army.

Remember that, if caught in a fierce fire, all petrol-driven vehicles would explode and burn. Ammunition would go up too. The enemy would lose all transport, all ammunition, all supplies. He would be finished.

There are numbers of areas in Australia where the enemy could be lured, or himself might march, into a patch of territory which could be fired without setting the whole country ablaze. Whoever organized it though would have to know that none of our own troops were in the vicinity. And he would have to be a bushman who knew what he was doing, otherwise his own men might be caught in his trap. To use such a weapon efficiently does not mean simply setting a match to a tuft of dry grass. It means a knowledge of the season, of the wind, of the quantity and state of grass and timber; a knowledge of the vagaries of fire, of the weather conditions, of distance and time; a knowledge of exactly where the enemy are and where they could be in an hour's time, and of other considerations besides. Then, under favourable circumstances, fire could be turned into a terrible weapon.

Take advantage of everything the country offers, anything at all that 'You could turn against the enemy: fire, water, hunger, bog, heat, thirst, distances, exhaustion, quicksands.

Remember too that you are fighting in the one way that will beat the enemy - defence but particularly offence in depth. If the enemy pushes you back or brushes you aside, it does not mean he is winning. Behind you for many miles are regular troops or other guerrilla bands, all operating, or eagerly awaiting the chance to operate, against the invader. The deeper into the country the enemy penetrates the harder he will be forced to fight; the more numerous and greater the difficulties and foes he must overcome, the greater and more pressing his difficulties of supply, transport, communication. With every mile he travels he will not only be forced to fight his front, but his ever lengthening flanks and rear. That fact gives you added confidence.

You can strike him in front, flanks, or rear. When forced to retire, it does not matter a great deal for you have plenty of country to retire to. Others will be tackling him while you reorganize or make fresh plans. The farther inland he advances the more open his flank become; you can again advance, pass his spearheads or let them pass you then get right to his rear. Thus he will be forced to continuously fight against offence and defence in depth.

Imagine the terrific and ever-growing strain upon his land, air and sea communications! That is why it is so important you should make him waste every bullet possible, for every bullet to him will count. Make him use up his petrol; destroy every truck - everything of his - you possibly can.

For he must replace everything. Offence and defence in depth is the only method so far found of defeating the Panzers, those extremely powerful, fast-moving mechanized spearheads of the enemy. It was the people's army in Spain that solved the problem. They found that fortified entrenchments and strongly held stationary positions could not stop these speeding masses of mechanism. So the people fought, wherever they were, with any weapon that came to hand; even with iron bars, picks, clubs, old shot-guns, knives, and pistols. Until they could take weapons from the fallen enemy.

It was fearfully costly to the guerrillas. But it stopped the onrush of the mechanized spearheads; because they had to fight continuously no matter how far they went, or where they went. There was always resistance of some sort to be overcome. The bullets of snipers constantly took toll. Bridges were blown up or burned, and had to be repaired under fire. Roads were blocked. People hurled grenades and bottles of petrol at them; threw petrol-soaked blankets on to tanks. They put a tank's machine-guns out of action by hitting the muzzles with hammers, then hammered the tank until the deafening vibrations forced the crew to crawl out. Constantly, the men of the spearheads were falling; a lorry here and there was blown up; occasionally a tank blundered into a trap and was destroyed. Convoys bringing petrol and supplies to the spearheads were ambushed, and always some of the precious supplies went up in smoke. This defence in depth, this people's army slowed up the mechanized spearheads and finally brought them to a halt.

China has fought thus these five years past. When we think what the .Japanese in the short space of five months have done to the British and Netherlands Empires, we realize how successful the Chinese guerrillas have been.

And Russia, using similar tactics on a grand scale, brought the German mechanized rush at last to a halt.

Guerrilla fighting in Australia will be a bitter fight of offence with defence in depth. Far more offence than defence, I hope. For offence is the mighty weapon. Get in the king hit first; every time.

Remember this though: There is not the faintest need to be discouraged again and again. If your little individual group is pushed back or brushed aside, you will know that, besides the regular forces there will be hundreds of groups like yours, all doing their share of damage to the enemy.

Think then of every aid that nature, and your own invention and adaptibility, can bring to bear upon the enemy. No idea is ridiculous so long as there may be something in it that can be turned against the enemy. Remember also it is you who have to do it; you are sons of Australia, sons of the old pioneers. What a wonderful job they did by using their own wits, energy, an axe, and the natural resources of the country. It is your job now to defend that same country. Do as your grandfathers did. Use your wits and energy and the natural resources, the natural geography, of the country against the enemy. Don't forget that our greatest weapon is guts, mobile offence (and defence) in depth.

You will be interested in tanks. Here is a little about them. Every nation jealously guards its tank secrets. Also every nation is feverishly and constantly improving its designs. Hence it is practically impossible to give up-to-date capabilities of the very latest tank. But we can go fairly close to it.

An up-to-date tank possesses considerable speed, (35 miles an hour) weight, and great fighting strength. But this strength has certain limits. For instance: it is strictly confined within the tank; it is limited by the supplies it can carry; and, above all, by the endurance of the crew and their limited vision.

The tank's range of travel is limited by the natural obstacles it encounters; by necessary repairs, enemy opposition, fatigue of crew, and facilities of refuelling, re-oiling, water for radiator, ammunition, rest and food for the crew. Hence, although a fleet of tanks might make a strong rush ahead for say a hundred to one hundred and fifty miles, their race is finished should their supplies be blocked behind them. And, sooner or later, the crews must rest.

A tank is a mass of machinery. Every "breathing space" inside is occupied by the crew or some gadget or another. There is the locomotive machinery, devices for radio, optical instruments, mechanical gadgets in electricity for the revolving gun turrets, etc. According to whether it is a light tank, cruiser, medium or heavy, so it is armed with more or fewer machine-guns, with lighter or heavier guns. Some fifty-ton monsters are today carrying a 6-inch gun, with a lighter gun, a battery of machine-guns, and a flame-thrower. According to the size of the tank, also, is the thickness. of the armour plating: three inches, four inches even. No doubt tanks and armament will increase as time goes on.

According to the size of the tank so is the number of the crew. A small tank may carry only three men; a large tank half a dozen or more.

A vital point of interest to guerrilla fighters is the crew of a tank. They are vulnerable men. On a hot day they are imprisoned in a hell-hole, providing their tank be any but one of the latest types. Dust, smoke, heat, thirst, oil fumes, jerking and jolting of the tank, poor vision, uncertainty - all these are their enemies. Then you have to be reckoned with. Could you increase their difficulties? If so, you may bring that tank to a halt despite its armour plate, its guns and machine-guns, its grenades and tommy-guns. Just warm up that armour plate a bit more and you have them, for they are already half suffocating.

Picture what they must put up with, and you may get an idea for increasing their troubles when the time comes to stop a tank. That with its great strength and speed may be beyond you. Concentrate then on the men inside; they really are the vitals of the tank. The latest model tanks are much more comfortable for these are air conditioned. Even so there is a severe strain upon the crew.

Each man sits more or less cramped up on a small, hard seat with gadgets all round him. There is no room anywhere to swing a mouse, let alone a cat. It is a warm day, hot in¬side the tank; the air is stuffy and smelly with oil; the crew are thirsty and sweating. The tank moves off on a rough road. She lurches ahead, rocks and sways. The engine gets up a roar, and the tank's track is kicking up the devil of a row. Each man sits in his gloomy possy with the din in his ears, breathing an atmosphere that warns of a headache. The officer, with head and shoulders exposed through the turret hatch, is watching the way, while each of the crew is peering through slit or periscope. It's a dusty road and soon the dust obscures vision. Some of it penetrates the machine-gun slits and air-vents and does no good to the nostrils. An extra nasty lurch jerks the crew against their gadgets; the engine and track whines and roars. Expecting the enemy, the officer closes the turret hatch and sighs as he descends into the murky, smelly gloom. All apertures except the necessary vents are closed. The tank grows hotter inside, while vision grows less. The men are a bit. nervy. They expect shells, other tanks, mines, tank traps, anything. Anything might go off at any moment. The driver feels like a cat on hot bricks, with his eyes and mind glued to the road and the mines he may bump over at any moment.

When not in action the driver has a good view from an open armoured cowl. But in action, or when in fear of snipers, he closes the cowl. His vision is then limited to a periscope which reflects to a mirror on the dash-board.

The vizor is only a few inches square, of splinter proof glass fitted against a cushion. This cushion takes the shock of a bullet. But the bullet can blur the glass. The gunner immediately drops an armoured hood down over the glass. This hood protects him while he fits in a new glass. It only takes a few seconds but he is out of action for that time. That armoured hood in modern tanks is so fitted that it cannot be raised from outside, hence there is no chance in this type of tank of leaping upon it, opening the hood, and squeezing a grenade through the hole.

If you have neither trap nor ambush prepared, and a tank comes lumbering along out of grenade range, then shoot at the officer if he is looking out of the turret. He very likely will be if he is not expecting a bullet. Give him one. It lowers the morale of the crew inside if they see their officer's legs kicking away down there. He is a nuisance in other ways too, for there is no space to spare inside a tank for a dead man. All the time he's blocking the way to the gadgets that regulate the tank. All in all he's a most disheartening influence on the remainder of the crew, even though they feel very lonely without him - or with him.

Make sure of the officer then and fire at the driver's "window", while others of you aim for the firing-slits behind which the gunners are peering. Providing you are good shots and know where to aim you'll disturb their nerves at least. Even with rifle- fire alone you may be able to kill a man or two. Don't despise this, even though you have no hope of stopping the tank. Every man in a tank crew is highly trained. Each one of the enemy you put out of action is a valuable gain to us.

We will imagine your tank is coming along, only to meet a road block. The crew are forced to deviate. On some roads they could not; heavy timber and other natural obstacles would pull them up. But in this instance it is plain sailing for them across open country.

What an increase to the jolting and lurching immediately they leave the road! What a terrific strain both to crew and tank in the lurching down into and climbing up out of ruts and gullies! The innards of the crew are jolted and shaken, as well as the innards of the tank. Still more human energy would be used up, and more petrol, more oil. So that forcing tanks to leave a road, helps to make them a little more vulnerable. If tanks could be forced to go "bush" in various parts of Australia they would soon have to pull up through breakdowns in negotiating countless gullies and logs and similar natural obstacles. Nor could the crew inside "stand the pace" for long periods.

I cannot say what distance a recently built tank can travel without refuelling. Up to recently it was round about one hundred miles; very recent models may have a range of one hundred and fifty miles. Water then as now was needed, both for tank and crew-in the heat of action something over a gallon a day a man. Therefore put every obstacle you can in the way of enemy tanks obtaining water.

Ammunition is also a pressing necessity when tanks are kept continuously in action, for they carry less ammunition than is generally imagined. Think up ruses then to lure them into wasting their ammunition.

Hence, if tanks could be forced to operate off a road, not only would their advance be considerably slowed down but they must incur breakages; use up their reserves of petrol, water, ammunition, and the energy of the crew. Increase their wastages; add further to their difficulties, and their resistance must collapse if they could be deprived of further supply.

A tank can climb a steep bank or gradient; but, unless something phenomenal has been built within recent months it cannot climb a gradient steeper than forty-five degrees. And it could not negotiate that obstacle by a sideways climb, for in that event it would topple over.

So (unless there is a big surprise for us) there is a limit to what a tank can climb. Which is rather interesting because, although a large, powerful tank can climb a steep hill, it cannot climb a five-foot vertical step. That seems ridiculous in view of what we believe we know about tanks. Still such is a fact - always allowing for the possibility of some very recent invention. Imagine a solid stone or reinforced-cement wall five feet high. This wall is built so strongly that a tank cannot break it down. The tank comes to the wall, and stops,

The obstruction is vertical, and at such a height that the nose cannot negotiate it. The tank must back away and blow a grip in the wall with its gun. Or the crew must get out and build up a step before the wall, a log or two or bags of earth, anything for the track to grip so that the nose of the tank will be lifted an inch or two above the level of the wall.

It is the build of the tank and the vertical steepness that does it.

Even hard earth may stop a tank. It must be a vertical" bank, but need only be six or seven feet in height. Such a bank was capable of stopping a tank until very recently. If you dug a tank trap in hard earth say six feet deep and a tank fell in, it would bury its nose into the foot of the bank but could not climb, unless portion of the bank gave way and thus formed a "climbing cushion" for the track. But you would have to be there with grenades and fire-bottles otherwise the tank would soon surmount the obstacle. The crew would merely have to get to pick and shovel work and dig out a bit of a ramp, when up and away would lumber the tank.

However, very recent tanks are fitted with "teeth" that would bite into such a wall and pull down the earth so that the track could grip. But production problems may be such that the enemy may not be a hundred per cent equipped with these.

There's another interesting fact to be considered if you thought of digging such a trap. Say you decided to dig the trap six or seven feet deep, and of such a width that the nose of the tank would dig in under the bottom of the opposite wall and the tail rest on the top of the wall behind.

The tank would certainly be in an awkward position. But it could get out; either by cutting down the opposing bank or, beforehand, throwing bundles of sticks at its foot. These bundles give the track the necessary grip to lift the nose up to the level of the pit - and away lumbers your tank.

If you could think out some little dodge to destroy the track of a tank, you would immediately render the tank immobile. That is the object of the grenadiers, to throw grenades under the tracks and thus break them. The tank could still fight, and if a comrade tank happened to be handy the cripple could be towed away, and keep fighting while being towed. But if there was no other tank, you would have the chance to do your stuff.

A bunch of four or six hand grenades tied together and thrown under the caterpiller tracks will give even a heavy tank a very nasty jolt, and very probably break the track.

Tanks have been skittled on other fronts by a guerrilla lobbing a grenade into the open turret hatch. Or by dropping it down the hatch from an overhanging tree-branch as the tank passed below. But such are "lucky shots".

Still, the chance may come to you. The morale effect, alone, of an unexpected grenade does no good to the nerves of the crew.

Over large areas in Australia there still remains much timber, varying in density according to the locality. This fact, where we can force enemy tanks off the roads, will be decidedly against the tank. I am not referring to areas where the timber is so thick that no tank could negotiate it, but to lighter timbered areas.

A tank can go through the wall of an ordinary house. A telegraph-pole means nothing to it. Neither does an ordinary tree, it can push it down. It can lumber across fallen logs. But (note the little fact) a tank, like a motor-car, has a certain clearance, say two feet. Therefore if a large log blocks its path it must negotiate so as to lumber over that log at an angle. Where there is standing timber with logs lying about, tanks would find the combination fearfully costly in petrol, man-energy, and wear and tear, in addition to very considerable delay. A tank might push down one or two trees and climb on over them, *but* it is not going to push down all the trees. And if it gets into a fix for manoeuvring room and finds a large log jammed under its belly, it is in for a torrid time, even if you do not set the bush on fire. You would be under ample cover all around the helpless tank. It should be a fairly simple matter to smoke the crew out. 1f a dozen tanks were caught like that, then fire the bush around them: a light fire with plenty of smoke; there would be no need to set the bush ablaze.

Another point about a tank is that its field of fire is limited. The turret regulates the angle to which the gun-muzzle can be swung, raised, or lowered. Machine-guns command a wider range of fire although their traverse is limited to about thirty degrees. This explains why guerrillas have been able to get close enough to wreck a tank with grenades. On such occasions they got right up against the sides of the tanks. The machine-gun muzzles could not then be pointed down sufficiently low to "get them". I've a suspicion it may not be quite so easy with the very modern tanks. Also there are such things as tommy-guns in the hands of the crew. Almost certainly provision will have been made for the use of grenades or gas bombs from within the tank. Even so, the field of fire available to the crew would be limited. Smoke or dust would blind them.

It takes a fair quantity of water, even up to a depth of about five feet, to stop a big, modern tank. At the same time, don't be sure that a deeper, but narrower creek will always stop a tank. Providing the bed of the creek is fairly level and the ascent on the bank is easy the tank may possibly get through. If it charges at full speed, its weight and speed displaces. the water to such an extent that the tank may get through. The stream must be narrow, however.

Tanks when travelling along a good road can shoot with fair accuracy. But it is a different matter when they are off the road; the rougher the country the more inaccurate their shooting. They must pull up to use their guns efficiently, and immediately they come to a halt they become good targets for artillery. At the same time a stationary tank can put in very good shooting at a moving target. The gunners in general, to allow for the oscillating action in tanks, have adopted the "roll-yaw-pitch" method of the navy gunners. Against aircraft, modern tanks use guns which act on a swivel turret.

Everything being taken into consideration, the tank

although a very formidable weapon, is weakened by strict limitations. These limitations enable you to plan with some chance of successfully trapping or ambushing a tank. Again, knowing these weaknesses of a tank, you can often defy it. For instance, if your possy is so placed that the tank cannot get at you, then you can tell it to go to blazes. The defiance of a bad boy to a policeman is good for the morale of the boy - so long as the policeman cannot get at him.

I believe we will make a lot of use of smoke in our tank warfare. None of the crew can see if the tank be travelling through smoke: Further, the smoke added to the heat within the tank' is suffocating.

All modern tanks carry several mortars, with smoke as well as H.E. shells. Tanks are increasingly using smoke in offence and defence. If they are attacking and are being attacked by anti-tank guns they throw out a screen of smoke before and around them.

This disguises their movements, and also blots them out of the view of the anti-tank gunners. It certainly also blinds the tanks. But the big advantage, apart from dodging anti-aircraft shells is that under cover of the smoke they can advance and change direction and he upon their enemy before they are aware of it. Although their fire is quite ineffectual while they are in their own smoke still when they emerge they are right upon their enemy. And then it is a bit late to do much about it.

It may be a point of interest to learn that anti-tank gunners also now use smoke. The flash from an anti-tank gun is very noticeable so the gunners at times use smoke bombs to help hide their position.

If snipers are expected at any time a tank is forced to keep closed. And this increases the difficulties of the crew, besides adding to their discomfiture.

Remember that tanks become immobile without fuel. Destroy the lorries which carry their fuel supplies and you anchor the tanks. This gives artillery a chance to come up and destroy them.

The commander peers through a little vizor, he has a periscope also. He is in touch by wireless and telephone with the squadron commander, by flag and Morse signal when it is not advisable to use wireless. He speaks to his own crew by telephone.

Difficulties of invasion have now been increased for hostile armies by the mere fact that they must have tanks. The landing of these tanks is the difficulty. If the enemy can capture and hold a port, then well and good. But the taking of a port is a major operation which might easily fail if the other side determinedly defended it.

This means landing on a beach. If the beach is defended then it is a task of great difficulty to land heavy tanks, particularly under heavy fire.

To cross a deep stream tanks must wait until the engineers build a pontoon. The nations are experimenting with amphibian tanks to obviate this delay. But so far no amphibian is anything like so powerful as a tank built for land use alone. Also, it takes comparatively little to stop an amphibian whether in the way of natural obstacles or shells compared to what it does to stop a tank. Perhaps an army may seek to use a few amphibians for the purpose of crossing a stream and seize a bridgehead on the opposite bank until the engineers can build a pontoon. Otherwise the amphibian is not nearly so far developed as the land tank.

And tackle the tank at night. It is then at a very great disadvantage. It cannot "see". Also, the crew will be tired. You have very little to fear from it, as compared to the daytime. Plan to surprise the crew. They will be out of their "shell", sleeping in the fresh, open air. The turret hatch will be open. In fact, providing the enemy are unsuspecting, everything should be open to your attack. Even if the crews scrambled back into the tank or tanks, still they could not see.

Always remember that limited field of vision, either by day or night. The field of vision of an enemy tank crew in the Australian bush by day would be considerably limited. Tremendously so if you had the chance of making them work through smoke.

The effect of smoke - oil tanks bombed by the Japanese in Darwin 1942.

3

Attacking that Enemy Aerodrome

DO not forget that a modern army marches and fights on its petrol. Petrol will burn. Burn it.

You help your country; help your army; help yourself with every tin of enemy petrol you destroy, every bomb, ammunition box, shell, cartridge, or bag of food you deny to the enemy. Destroy all you possibly can. Make them waste all you can.

The enemy may land in Australia with an army on the east coast, another on the west, perhaps another to the north. Some of his forces will possibly succeed in advancing. That is your chance. Get to his flank and rear. Burn his lorries laden with supplies; find and burn his supply-dumps; his workshops, bush dromes, grounded planes-everything that is his. Fire is a terrible weapon. It is cheap. Use it.

He must prepare hidden bases of supplies. He will make dumps in the shelter of the bush to defeat the prying eyes of our planes. Your scouts must   find those supplies. They must also find enemy dromes to his rear. One of his very first jobs will be to establish land dromes if he can. And he will conceal them to the very best of his ability. He'll use dummy planes, dummy hangars; will camouflage his own planes and any building activities.

Now, such a field base would almost certainly contain more men than you could muster. There would be the guard, the ground staff, and the actual flying men. There would be defences in the way of machine-guns and tommy-guns, not so much perhaps against ground troops, as against attack from the air. Such a drome would be well behind its own lines and not expecting ground attack. But the Australian guerrilla must at all times be prepared to strike from anywhere, especially from unexpected quarters and at unexpected times. If you rushed in and attacked such a drome you might be wiped out. Whereas if you first thought out a shrewd little plan, and acted on it efficiently, you would do a lot of damage. But think - could you do a hundred per cent damage? That is what you should aim to do *every* time - total damage.

Think it out first. Your scouts have located an enemy drome being formed in a pocket of forest country. There are only half a dozen planes there now. But through the forest country enemy lorries are winding, bringing supplies: tucker, bombs, petrol, oil, repair outfits. On the rough bush drome men are busy clearing away stumps, camouflaging planes, preparing secure possies for petrol, food, and aerial bombs. Within a few days there will be more than half a dozen enemy planes there; there will be a score.

Very well then, you will be doing a far better job by destroying twenty enemy planes, than by taking the chance of destroying half a dozen or so. You decide to carry out the job one hundred per cent. Your scouts, well hidden, watch that rapidly forming enemy drome, day and night. They must find out the number of planes which day by day are parked on that area, and just where each plane is. They must find just where the petrol supplies are going to be kept; whether stacked in a camouflaged pit, or just rolled together in some handy spot under shelter of camouflaging trees. Your scouts must learn exactly where the aerial bombs are to be stacked, where the ground staff sleep, where the pilots and observers and headquarters are. camped. And the approach from the nearest main road, or their made route from the coast, where lorries with supplies may at any moment be expected to come along. Your scouts must further learn, by spying at sundown, night-time, and dawn, where the outposts or sentries are placed.

In due course you learn very closely the numbers, positions, and habits of the enemy. You choose the time when you feel your numbers and resources can best tackle the greatest number of planes which the enemy will one night have upon the ground.

If you cannot now get further help from the military, or from neighbouring guerrilla bands, your objective is sudden and well-planned surprise to cause the greatest possible damage to the enemy machines and the supply depots, and the heaviest casualties to the personnel. Then you vanish.

You attack just after the dead of night, preferably in those small hours when sleep is soundest. Your scouts have told you exactly where the sentries are.

Now, be careful. There will possibly be a ring of sentry posts, but they will be widely spaced.

If you could quietly kill one outpost you could creep through to your agreed-upon positions. Then at a given signal spring to your separate jobs: some to destroy the planes on the ground by bomb and fire (Molotov cocktails are good); some to puncture and fire the petrol-dump; some. to fire the aerial bombs; some to rush the enemy machine-guns and turn them on the enemy personnel tumbling from sleep.

But, can you quietly overpower the sentries? Believe me, it is a very difficult job. The sentry who feels a sudden bayonet or knife in his back screams, yells, or gasps. He certainly jumps, and the noise wakens his comrades. If you bash him on the head and don't do it scientifically his yell will fill the skies. If he is seized from behind and feels suffocation upon him he struggles violently. No man likes to die violently in the night. If he cannot help it, he makes all the row he can. If he is throttled and lifted off his feet at the same time, there must be some sound or movement by which the men sleeping beside him are awakened.

You certainly might put one sentry quietly to sleep. But what about his cobbers? If say, three to five men on an out-post are killed noiselessly it is an uncanny job. If you can manage it, you are . . . A lot would depend on how far away the outpost was from its fellow posts, and from the men who were depending upon its watchfulness. The farther away of course the better chance you would have, for a bit of a gurgle now and then and a kick or two wouldn't be heard so very far away.

If you are doubtful about it, realize that the sentries will be widely spaced and it may be quite possible for all of you to crawl unseen between two, or more of the posts. The scouts should look to this beforehand. It is not so much the killing of sentries that is your first vital objective, it is to be as near as possible to those enemy planes before the alarm is given. The closer you are the easier for you to destroy them. It is a fact that it may be more to your advantage to crawl unnoticed between two outposts, than take the big risk of killing sentries. While the sentries are there "All's well" to the enemy. A relief can take over from those sentries, an officer can visit them and still "All's well". But if the sentries are found dead . . .

It is up to you to crawl between them. Those of you whose objective is the planes, concentrate towards the planes. The petrol party worms its way towards the petrol-dump. Each group to its objective. Those who are to turn the enemy machine- and tommy-guns on the ground staff when they awaken must creep towards the guns with a scout who knows just where they are. And so on. It depends on the number of your men; if very limited, you can only concentrate on the planes. Destroy every machine possible then vanish in the confusion. But if your numbers are large enough, try to clean up the whole show.

You are all keyed up. You know what everyone of you must do in teamwork. If your numbers are sufficiently large, it mar be advisable to detail men to watch the sentries while the rest of you are crawling past them. If you get through, all will be well for the time being. Each can get to his job and have time to wait there until all are ready. But if a sentry suddenly "wakes up" and gives the alarm, the watchers rush on the sentries and bayonet them.

That is actually the signal; the enemy makes the signal for attack himself. All the rest of you instantly leap to your jobs. Those for the planes, rush them and throw grenades and Molotovs into them. Those who are to spray the personnel with tommy-guns and grenades do so, before the sleepers can tumble from their blankets. And so on.

If no such alarm rang out, you would have made a good start with your job before the enemy woke up to it.

With such a planned surprise, a small number of you could do great damage to a considerable number of the enemy. They would awake with their blankets being ripped by bullets, grenades bursting amongst them, their planes and petrol-supplies bursting into flames. They would be frantic for the time being whereas you each would swiftly be doing 'Your allotted job. Then all would melt into the bush.

If you were strong enough you could remain masters of the situation.

In such a case, mop up all survivors before dawn if possible. Complete the wreckage. If all was clear at dawn hastily camouflage the wreckage, unless there still was so much smoke that it must inevitably put you away. If not, put the place "in order"; enjoy a good feed at the enemy's expense; await any planes that may come down on the drome; and have a reception party ready to greet any enemy supply-trucks that presently will come bowling along.

You should collect quite a lot of loot, not forgetting valuable papers.

Such a simple plan under the conditions outlined would almost certainly succeed if carried out efficiently and boldly. The matter most vital for your consideration beforehand would be the information. And this would depend on the scouts.

We'll imagine, now, that you have located a larger drome, much more widely spaced: more planes, more men, more defences. You doubt if your small numbers can tackle it without being cut to pieces. You cannot get in touch with our Military to send over planes to bomb the place, or for some reason it is not in our Military's power to do so at the time. Yet it is vitally necessary that as many as possible of those enemy planes shall be destroyed as quickly as possible. You are all game as pisants to tackle this apparently hopeless job. Your only chance of success is by using your wits.

Your problem boils down to this, and to this only: How to divert the attention of the enemy from their precious machines; also to bring as many as possible of them rushing to as distant a point as possible from those machines.

See if you can do it.

The petrol stack, dump, tanks, or supply, are certain to be away to the side or end of the drome (perhaps in a camouflaged pit), some distance from the machines anyway, for the enemy would seek to guard against accident, setting the petrol on fire. There may be only one petrol-dump or several scattered around the drome; or there may be scattered "groups" of petrol-supply. That is for the scouts to find out.

Now, if you can reach petrol-supply (or any supplies) and fire it (as "accidentally" as possible) the enemy would hasten to put out the fire before it starts petrol and perhaps bombs exploding, and thus spread to the machines.

Then would be your chance. You could rise from the ground, race for the machines, and do 'Your job while the enemy's attention was distracted. When all damage possible had been done, scatter into the bush and meet at an agreed. on spot later on.

The getaway of every operation must be planned beforehand, just as well as the attack.

According to local conditions and circumstances though you might conclude a successful little raid by leaving a "string in your tail". Imagine you attack an enemy post - any position manned by any manner of enemy forces. There is a bush track leading to and from it, or the enemy have formed their own track. You decide that, if all goes well you will not make your getaway by fading into the bush, but by ostentatiously clearing away hell for leather back down this track.

The disorganized, excited, and very angry enemy will pursue. Come after you full pelt, on foot, motor-cycle, armoured car, anything handy that they can leap into. At some narrow part of the track their leaders crash into a taut wire; those following pile up; grenades suddenly crash amongst them; rifles and tommy-guns are pouring bullets into them. Possibly a land-mine explodes under a lorry load of pursuers.

It is a nice little "string in the tail".

Throughout all your activities your scouts, when they see any enemy mechanized force, whether of the ground or air, must learn as much about it as the enemy officer in charge. That is, they must watch it until they know its strength; above all, until they are convinced that it is mobile or not; with petrol or out of petrol.

That point can be easily overlooked, although it is so important. It may easily mean a victory to you.

If a string of lorries or tanks is only crawling along, it may be because petrol is running short. Presently they will be forced to pull up. Probably they will turn off into the bush to seek cover against aerial observation, and wait there until petrol-supplies come by air or road. They may have to wait hours, possibly days. During all that time they will be much easier prey. Also, knowing that supplies must come, you could ambush the supply convoy.

If the enemy have formed a bush drome the scouts must notice how many planes are active; they must closely watch the actions of the ground staff, the jobs they are doing; and whether the pilots appear to be in immediate flying trim or not. If a number of planes are grounded and there's no sign of movement those planes may be inactive because the drome is short of petrol. If so, this is your chance. The machines cannot move. They may have far more enemy defending them than you can tackle. Still, you can fire the bush around them. If you can get in touch with our Military, our planes would come along and fix the job of destruction. But you may not be able to get word through in time. Then, can you fire the bush so that the fire will destroy the vehicles?

If conditions are not suitable, can you start fires out from the enemy a bit so that numbers of them will be forced to put out these fires, and so give you a chance to damage at least some of the machines? Or can you start a number of threatening fires, while at the same moment opening up on them with rifles and machine-guns? Caught between fire and bullets the enemy might be forced, in a hurry, to both beat off your attack and beat out the fires. In this dispersal of their numbers, interests, and activities a few of you might get near enough to the machines to do the job. The smoke would hide your weakness in numbers as well as your individual actions. Thus you might destroy a number of valuable enemy machines simply because, in the first place, your scout had noticed that those machines were temporarily immobilized through lack of petrol. Keen and efficient observation is priceless in all guerrilla operations; just as in every guerrilla action that we have discussed surprise and boldness is absolutely necessary. First the information upon which you act must be as complete and accurate as humanly possible. Then form your plan, not forgetting the getaway. Then act. The complete surprise must be carried out very boldly. Let them have all you've got, especially in those first few minutes. A well-planned surprise attack carried out boldly can, within the first few minutes, bring an enemy's numbers down to the level of your own, or lower, besides completely demoralizing them. It depends on your weapons and on how efficiently you use them of course, and on the enemy's numbers, disposition, and leadership. So, shoot their officers - quick.

Each man must fully understand his own individual job, dovetailing in team-work: the scouts, the men who will attend to the sentries, the men who will crawl between the lines to fire the machines, petrol, dumps, supplies, as the case may be, the men who will cover the enemy personnel, The principles are just the same if you attack an infantry post or garrison or column. Remember, too, that machine-guns suddenly and well directed on troops guarding an enemy position or activities tend to keep their heads down, not only causing them casualties but keeping them pinned there, thus allowing time for your mates to carry on with their work of destruction.

Don't forget in every action, small or large, to be on the look out for enemy maps and papers. These may prove of the greatest value to our High Command. If you raid a wireless station or ambush a solitary motor-cyclist, search it or him carefully for documents. Detail a man whose special job it will be in a raid to rush the O.C. enemy dug-out, wireless, or other similar objective to scoop up all papers. You never can tell when you may thus secure information vital to our regular forces.

Nabou Fujita, who flew a seaplane over Darwin prior to the initial raid, to prepare for the mass bombing of the Darwin port.

4
Wreck that Train!

ALWAYS plan to produce one hundred per cent results. Don't be content with half a loaf.

Say you have decided to blow up a bridge on a road constantly used by the enemy. If you blew up that bridge and got clear away, it would be a good job for you. It would delay supplies going to the enemy's front line, besides causing him inconvenience and waste of time, materials and labour necessary to repair the bridge. But could you cause him considerably more damage? Could you destroy that bridge at a time when a number of lorries were crossing it? And cause still more damage later on?

All that could be done in this way: Say you blew up the bridge while several lorries were crossing it. Several of the following lorries might pile up also; it would depend on their speed and the distance between them. The remainder would pull up and look to their defence. If the position and circumstances were that you could remain hidden up to that time, you could open up and cause a lot more damage. But think a moment. If you held your fire the enemy, recovering from their shock, would dismount and examine the damage. It would have to be repaired, and with all speed, for following lorries would be brought to the halt by a broken bridge.

Engineers and pioneers must be brought along at the quick and lively to repair that bridge. Then, while the place swarmed with dismounted men, would be the time to open fire.

Besides causing considerably more damage, you would destroy a lot more lorries and probably get away with many others. That would depend, of course, on circumstances, on position, and cover. If you had ample natural cover, and if circumstances allowed you to blow up the bridge or fire it in such a fashion as to give the enemy the impression it was a "hit and run" guerrilla job, they would concentrate on repairing the bridge. You could then wait for your target - a long line of stationary lorries with all hands in the open helping to repair the bridge. Sudden concentrated fire under such circumstances would be devastating.

It could be a case in which, after blowing up the bridge, lack of cover or force of circumstances compelled you to retire. But, you could come back at night when they were repairing the bridge.

The above illustration is given you to think over. Always first plan to do the greatest possible damage. Then act.

Remember, also, that by waiting a while you might do greater damage. We'll imagine that bridge again. The enemy have but very recently captured the road. Their front line position is insecure. So far as that bridge away behind them is concerned they have a guard upon it.

If you tackled the bridge now you must tackle a guard nervous and alert. The traffic using that road would also be alert. It would mean a sharp fight whilst the greatest damage you might do would be to destroy the bridge.

If the matter is urgent you will attempt to do so. If not, then in a few days the enemy may have consolidated his advanced position, and by so doing shows that he believes the road is his. If so, the guards will have been moved on while the traffic using the road, believing that immediate danger has past will lose their caution. And that would give you the chance to do a big job.

Always try to create panic. The more unsuspecting an enemy the more liable they are to panic under the shock of surprise. And panic spreads like wildfire. Even trained troops are liable to a panic fit if the shock is. unexpected enough and sustained long enough. Then, it is amazing what may happen. You could find yourself left victor with spoils so great that you would be unable to rush it all away. Do not linger, though. Make hay while the sun shines. Rush the job through before the panic has time to fizzle out and the scattered enemy reform for their revenge.

Whenever you desire to create panic like that always have sharpshooters detailed at strategic points whose job it is instantly to pick off any enemy officers who try to allay panic. It will be stiff luck for the officers, but it means you or them.

A panicky enemy who sees a cool-headed officer picked off immediately he shouts his commands becomes more panicky still. There is no holding them then.

If you've captured a trench mortar or two, lob their shells into any group that clusters to make a stand. The explosions will blow the groups away and panic reigns supreme. At a safe distance lob grenades (mortar shells if possible) into any lorry that appears loaded with ammunition. The resultant explosion will blow the enemy morale sky high.

It is not always necessary to blow up a bridge to stop a convoy so that you can get a crack at them. A few suggestions have been made already in *Guerrilla Tactics.* Perhaps the enemy are using an open stretch of road devoid of cover for your work. That, however, will be against the enemy too, for he will risk being spotted by our aircraft. So he will use that road at night, and night is the friend of the guerrilla. You do not need a broken bridge to bring the enemy to a halt. A few logs and rocks across some strategical portion of the road will do it; for then the convoy must come to a halt, and the drivers and their offsiders must climb down to remove the obstructions. Then you turn your tommy-guns on them and lob grenades into the lorries. And then, too, your trained men leap up into the drivers' seals and get away with the spoil.

Any country which is partly occupied in force by an enemy temporarily loses some of its railway system, which the enemy immediately turns to his own advantage. We must do all in our power to prevent the enemy gaining even a few of our branch lines. But, as there is no foretelling the fortunes of war, he may do so. Guerrilla railway wreckers must then do their utmost to smash the enemy railway system. Doubtless railway men will be detailed for this job; men whose experience will immediately tell them what to do in connexion with railway wreckage and sabotage. There will be troop, supply, and ammunition trains to wreck; rolling stock and everything that goes with a railway system.

Railway men will be able to do the job much quicker and more efficiently than other men for they will know the vital parts to wreck. But in case circumstances give you the opportunity of a bit of railway wrecking we'd better glance at it. First, what will be up against you? The line will be guarded. But railway lines are long; and an invading force will need every available man for its front line and lines of communication. There will be none too many men to spare for the guarding of railway lines. In all probability the guards will be widely spaced.

If you can do no greater damage, snipe them. It will be a much more exciting form of wallaby shooting than you've been used to. Remember that each "wallaby" you bring down must be replaced by two more wallabies, for once the line is sabotaged, the enemy will not dare leave it inefficiently guarded. Every man that you can force the enemy to put on to that line means he is a man short in his front line. Which means you are saving good Australian lives.

So snipe the guards. There are numerous Australian railways so wonderfully protected by natural cover that they would be a sniper's dream. Let the driver have it too when a train is passing by, on the off chance that you may cause a runaway train. Scout around the steep embankments, cuttings and tunnel entrances for a possy from which you can lob a grenade into the cab of an engine, or on to the trucks of an ammunition or petrol train. Plan it out. A couple of men could cause damage that would surprise you nearly as much as it would the enemy.

A line might be guarded by the enemy in various ways: with a few men at each strategic point, such as tunnel or bridge; with a man or a couple of men at considerable distances apart along the line; or with an armoured train or car - which would be dangerous only while it was within firing range. Probably a line would be guarded by a few men here and there; or with a pilot car travelling ahead of each train (once the line had been interfered with). It would not matter to the sniper, for the more men on the line the more targets for him; he would not have to travel so far for his shooting.

If your numbers are sufficient to attempt something really serious against the enemy, don't wake them up boy sniping first. Leave the sniping until after you have done the bigger job. Your scouts should watch the line until they obtain in. formation as to how it is guarded; and ascertain that portion of the line within your sector which would cause the worst damage to the enemy if wrecked.

It may be a tunnel. You must consider then the times when the enemy's main trains pass through that tunnel, and whether a pilot engine or car travels ahead. Also, such details as the time factor, and whether it would cause the greatest damage to blow up the tunnel itself down about its middle, or to place mines there, or tamper with the line there so that a train would wreck itself deep within the tunnel and cause an explosion in the roof that would bring portion of the roof down on it. There are quite a number of details to consider. Remember, your first chance will be your best. Mess up that opportunity and you won't get another, not just here anyway.

You must make a thorough job of that tunnel; cause a blockage that will last for days.

You must consider then the pilot train. This must be allowed to pass through as if all was well. Which means that your explosives must be of a type that you can fire at will. Otherwise you would only catch the pilot engine. You want to catch a heavily-loaded ammunition train within that tunnel. Wreck such a train deep within there and the fire if not the first explosions should explode the ammunition trucks and bring down the whole tunnel. It would take a week or more to repair it, while the blockage to the enemy's supply-line might well mean to them the loss of a battle. As to picking 'Your train that depends to a large extent on your scout's information, on whether the enemy have developed a system of supply, and on whether you are not detected.

The tunnel will probably be guarded. You must learn the position and hours of duty of these guards, their system of reliefs, and their system of signalling oncoming trains. The guards may have to report to their rail headquarters every hour, or two hours, or so. You must know about that. It would be useless to knock out the guards, then get busy laying your mines, if those guards were expected to report at frequent intervals. You must be capable of using their signal system, if any. If they are not in touch with rail headquarters and merely signal "All clear!" and so on to oncoming trains, all is simple. You do the signalling for them when the time comes. Of course you must look exactly like them, or the men in the passing train will "take a tumble".

You must also find out when the guards are relieved every twelve hours, every twenty-four, etc. You do not want a new relief to come along just when you have half completed your job.

All these matters of detail are vital points. For if even one detail is not considered the whole plan may well be wrecked. When you knock out the guard you must have thought out beforehand just exactly how you are to carry on impersonating him. Not only because of signalling or recognition or a casual glance when trains may pass by, but because an enemy railway line walking patrol or a hand-car patrol may come along. In that case if your bogus guard appears all it should, the rest of you can easily scupper the newcomers when they come unsuspecting into your net. Your guard must also be on the alert to deceive any low-flying patrol plane that skims overhead. All manner of detail must be carefully planned out. Then, except for some unforeseen happening, the plan will be a big success. You will cause the enemy great damage and will save the lives of men holding our distant front line.

As to the actual wrecking of tunnel and train this is simple enough if you have a man or men in your band used to explosives. It would be difficult but not impossible to make a thorough job otherwise. You want the rails to blow up at the same time as the roof comes down before and on top of the engine. The resulting crash and jar would probably set the ammunition off. But, to be sure, you'd need further mines or other explosive powers farther back to catch under the trucks. If you did the job to the engine and wrecked the trucks without their contents exploding, then you'd have to fire the trucks.

You can smash the rail by gelignite, dynamite or gun. cotton and detonators, the train setting off the explosion or else fire it boy electric battery. This would depend on whether the pilot engine travelled far enough ahead of you to lay the trap then get out of the tunnel before the oncoming train. There are various ways. Gelignite and instantaneous fuse would bring down the roof, just when you wanted it. Or landmines could be used. The best way of course would be to fire the mines by wire, if you had the simple apparatus. Then, in perfect safety you could let the mines go when you had the train just where you wanted it.

To make a really good job you'd need a man in your band who was used to demolition work by explosives. If you have no such man but find a good possy to blow up, quickly get in touch with the nearest Military, or your neighbouring guerrilla. Every guerrilla band should have at least one man who is familiar with explosives.

Another vital spot on any railway line is a bridge supporting it. Completely wreck this and no train can pass until the bridge is repaired. The ideal way in warfare is to. blow up the bridge with a train passing over it. The speed and weight of the train complete the wreckage of the arches or other supports of the bridge. Conditions would probably be similar to wrecking a tunnel-with this big difference: you can work in a tunnel at daytime, but only under very favourable conditions he secure from observation in daytime when working upon a bridge. Also, the getaway must be considered. A tunnel would be in country offering abundant cover and a clear getaway, A bridge might, or might not be in mountain country. You probably would have to work entirely under cover of night.

To derail a train is not very difficult. It causes the enemy damage, and holds him up until the wreckage is cleared and the line repaired. A curve with a steep fall on one side is a nasty place. A few sticks of gelignite, or dynamite with detonator cap on the rail will send the train plunging overboard. Also, the line itself can be tampered with, without explosives.

Trains must have water. Some watering-places are in lonely localities. Such might easily offer a chance to capture and destroy a supply-train, There would only be the guards to overpower. A few well-placed grenades as the train came to a halt would probably do the job.

If ever you lay hands on an undamaged engine send it full pelt back along the line towards the enemy. The enemy is sure to want their engine, but they won't want it back the way it comes. If you could pack it with shells from captured trucks then when it reached the enemy it would arrive home with a bang.

Remember the landslide. Granted the explosives, a little time, experience, and elbow grease it can be man-made, There are fairly numerous places along our mountain railways where nature has laid the foundations for a landslide. Upset these foundations and the thousands of tons of rock and debris that would come pouring down on to an oncoming train would be just too bad.

The first Japanese Zero downed near Darwin.

5

Trap Them! Bomb Them! Mine Them! Smash Them!

DON'T forget the little tricks. Small fish are sweet. In *Guerrilla Tactics* the point was made that the enemy are hungry; that nothing gives them greater pleasure than getting their teeth into our good tucker. Remember the farmhouse ambush, and the nice fat sheep innocently browsing, with invisible snipers shepherding them. Never forget that the enemy has an appetite. Although he is an uninvited guest, we're the host. Give him what he wants.

A few sheep browsing beside the road, a dozen chooks picking up a handful of wheat will stop a lorry load of troops. Try it. Hide by the roadside, with a scout a mile away up along the track. The scout signals when the enemy are coming. Then take your hungry chooks out of the bag, show them the wheat and scatter it out on to the road. Your machine-guns, your grenades, are all set.

When the enemy spot the chooks there'll be a loud "Hurrah!" in whatever language they bring to these shores. The lorry will pull up, the boys will hop out and, with a "Yakai!", run to surround the fowls. Which incidentally will scatter to the bushes.

Don't join in the hunt straightaway; wait until the driver also has joined in the chase. If he won't, shoot him. Then open up on the others. You'll have some fine sport. And a perfectly good lorry given in.

Even a hungry tank crew could not resist the sight of a dozen fat sheep quietly browsing a hundred yards or so off the road. You'll be on the opposite side but they won't know that- until too late.

Never forget the tucker-bag. Again and again you ought to get a good bag. Whether its nine or five miles from Gundagai, the goats will be there.

A pig will catch a dispatch rider. Try it out. He is travelling at forty, fifty, sixty, seventy miles an hour on his motorcycle. You can't stop him, and for some reason or another you're not going to stretch a wire across the road. But if you've got a hungry porker, sling him a bit of tucker just before the dispatch rider comes roaring along. From afar the rider sees that porker, makes a lightning calculation. All is clear. It will only take a minute to stop, catch or shoot him, sling him on the bike and away again.

He pulls up, dismounts, decides it will be quicker to shoot the porker, Let him; it's your porker anyway. He shoots and comes hurrying across to secure his prize. You shoot him; tit for tat. You get the dispatch rider, his dispatches, his high. powered motorcycle, and the porker.

Then there is the egg trick. The first thing a hungry man does when he sees one is to pick it up. Could you fill the inside of that egg with explosive so that when it is touched it goes - bang!

The enemy will need horses for transport. We must move all horses possible before the enemy can lay hands on them. Still, there is sure to be a stray or two about. You could make a trap with a neddy, but be sure that no harm could befall the old prado Have him browsing half a mile or so off the road. Two or three of the enemy are sure to come after him. But you are within fifty yards of him - with a rifle.

You could play numberless tricks like that, just to keep your wits in. What self-respecting enemy could ignore a fat gaggle of geese waddling away down the paddock, or the homestead ducks enjoying a bather at sunset? They'd make straight for the goose pond, the waterhole, or dam as the case may be. You'd be there too.

It's a real shame what you could do in this line if only 'You use your wits. Think out any trick to lure the enemy away from the road or convoy or position, into your trap. Many stratagems could be worked out, even by two or three guerrillas working on their own. As for a large guerrilla band, it is just as easy to lure a fair-sized body of men, as half a dozen or fewer. It all depends on the bait.

Lure larger units of the enemy deep into the bush; into country where you know every bridle-path, hill, creek and ravine, every patch of scrub, every waterhole.

Let there be an ambush awaiting them on every road, at every crossing, every bridge, every waterhole, above every cutting, around every bend, along every track, at every camp at nights. Attack them in the dead of night; attack them at dawn. Let them fear a sniper lurking behind every rock, and tree and bush, in every depression in the ground, away to their flanks, behind them, at their very feet. Shoot up aircraft on their very landing-grounds; ambush their staff cars; kill their dispatch riders; waylay their stragglers; kill their outposts. Throw grenades amongst them as they sleep at night; drop grenades from the trees and rocks above as they pass below. Throw grenades into their tents, their shelters; fill their dreams with explosions and sudden death. Bring creeping death to their sentries, hell to their every unit. Keep their nerves on edge, make them fear every rustle that may mean more than a snake in the grass, every echo, even laughter. Teach them to fear the dark roads at night, the innocent bush by day. Kill them before they reach the firing-line. If they kill one of our civilians kill a hundred of them. Derail supply-trains; alter switches on railway lines; play monkey tricks with signals. Alter road signs if they put up any. Thus you will send their transport and reinforcements on the wrong track, preferably into an ambush.

Be on guard against their own monkey tricks: frequently they sham death and lie in wait for you to walk up to take their arms.

The enemy will strain every nerve to get in touch with enemy subjects who may have escaped internment; these will certainly act as Fifth Columnists. If you catch any, of course shoot them. Keep your eyes and ears intensely alert for any sign of Fifth Column activity; learn or guess by what means it may be playing into the hands of the enemy, and then swiftly use your knowledge to our advantage. Give the enemy all the Fifth Column work he wants; play it off against him. If German, Italian, or any Fifth Columnists are working, then send them Fifth Column messages and information. Scrawl German and Italian "information" on rocks by the roadside, on culverts, anywhere a columnist may apparently write it for the enemy to see. If the enemy acts on it he runs into an ambush.

If you detect Fifth Column work actually being carried out, circumstances may be such that you immediately shoot the men responsible. Otherwise, think carefully before you make your presence known. We will imagine that your intelligence service has discovered a man or men who are using some system of signalling information to the enemy. Would it pay you better to let them work on, believing they were unobserved, until you could take over and do the, signalling? If so, you could get in touch with our Military and lure the enemy into a big trap indeed.

Remember that this vast scale war is a world war, planned by many of the cleverest war-brains in the world. It is a practical certainty that in our midst, trained enemy aliens are planted in strategic places, with hidden instruments ready to operate for an invading forces' benefit. If so, Fifth Column enemy activity of a trained and very dangerous type may become immediately active should an enemy land. Be on the look out for any sign of this subtle or apparently uncouth, such as the scribbling of a sign or message in charcoal on a rock, an arrow pointing on the ground, a tree lying at a road crossing, or anything that may be an agreed upon enemy sign, signal, or message.

If you could detect any form of such work, and quietly keep in the background, you could soon form a plan to "take over" at the right moment and cause the enemy a disaster.

Fight him with your wits. Your wits are a far more dangerous weapon than your skill and bodies. If you use your wits you may easily kill a hundred of the enemy. If you use your rifle only, you may only kill one.

Week by week you can produce similar results. Wits win every time.

Don't forget the booby traps if forced to abandon homestead or township. Be absolutely certain though that the enemy will swiftly be in the township and that our own troops have abandoned it. While there is time then put a booby trap at the front gate, in the cupboard, under the mat, in the kettle. When the enemy turns on a tap the whole business blows up, when he opens a door-bang! 'When an enemy pulls a chain-bang!

Booby traps are highly dangerous, but very simply arranged. You must use imagination and skill: imagination to set the trap where it will be least expected; skill in setting, so that it will explode to the "tick" yet will not blow you up in the setting.

For instance, most grenades explode a few seconds after a pin is removed. If this pin is *almost* removed then fixed in the *almost* position, the grenade will explode if it is the least bit disturbed. For that will mean displacing the pin to its limit. Just as with a loaded rifle, pull the trigger and the cartridge explodes. Look. on the grenade pin as the trigger. Some slight enemy movement must be the "pull" which explodes the grenade.

Pick on a cunning hiding-place for the grenade. Then *carefully almost* withdraw the pin. Let it rest against any obstruction which will hold it in that position. Previously tie a fine string to the pin and lead it away to some position where sooner or later an enemy must kick it. The thread then jerks the pin to its limit and the grenade explodes.

Such a string need not be visible. For instance, it could be laid under flooring boards, with its end tacked to the bottom of a leg of a chair. Immediately the chair is touched the trigger is pulled.

The grenade could be in the wall, or in the ceiling. The end of the string could be fastened to the bottom of a saucepan - to anything.

Such a trap (even a mine) does not necessarily need a string. For instance, take our Mills bomb. First, to throw it as a hand grenade: pull the ring, which pulls out the safety pin.

The lever then flies out, if allowed; but you hold the grenade with the lever held to it. When you throw the grenade. the lever flies out and thus ignites the fuse.

Now, if you get a glass jar, put the Mills inside and carefully manipulate it against the jar, then pull out the safety-pin, The lever immediately edges outward. But if you have the grenade so placed that the lever only edges out a certain distance then is jammed against the side of the jar, it will not explode. But it will explode if that jar is broken, for then the lever can extend to its full length.

Thus, if you buried such a jar in the road, merely dug a little hole then covered the jar with the dust, the first vehicle that passed over it or the first foot that trod upon and broke it!

To turn the jar into a mine, wire four or five grenades around it so that the one that explodes explodes the lot.

So, you see, lots of tricks can be played with our own, or enemy, grenades that explode per lever.

If the grenade was placed in a tin, say a canister labelled sugar or tea, and the safety-pin carefully withdrawn so that the outstretched lever was held by the side of the tin, then anyone who opened or touched that tin ... ! Or, under a loose board, the weight of a foot displacing the lever - "bang!"

With the jar trick, a grenade or a big mine could be exploded day or night, as required, by still a different method. But this is cumbersome in a manner of speaking. It needs considerably more time and manipulation, and may need the sacrificing of an enemy rifle. The principle is that a glass vessel can be broken by a bullet. As the safety-pin is previously withdrawn the lever flies out and the grenade explodes. If other grenades are heaped around, then the mine explodes. The advantage of this is that the grenade (or mine) can be sprung at any moment you believe it would do most damage -

and at a fairly considerable distance away.

You can place the mine anywhere, providing there is a clear field for the "firing" bullet. The bullet in this case is the trigger. Say 'You wish to mine a road over which you expect a column of enemy infantry to pass.

You dig a small hole where you decide to place the mine. In that hole you place something as nearly as possible the exact size of the glass or earthenware jar. Then retire to a position, off the road that any stray enemy is least likely to stumble upon. Right there, you fix up a vyce of some sort. Between two rocks perhaps, or two logs. To make quite sure, you fashion a vyce of stakes in the ground, which you can cover up later. You fix the rifle loosely in the vyce and aim it at the target - the dummy jar down on the road. You fire until you hit the target. Then clamp the rifle tight. Fire it again until the tightly clamped rifle is dead on the target. You now know that a bullet fired from that rifle must hit the jar.

Then you put in the mine (the jar and the grenades) exactly where the dummy was. Lightly cover the mine. Put a bullet in the rifle. When the time comes, you fire the rifle; the jar breaks and the mine explodes. You vanish away into the night with a flying start. The rifle is two hundred, three hundred, even four hundred yards away from the point of explosion. The enemy have no possible hope of catching you.

They would not dream how the mine had been fired, nor from what direction.

If you wish it, or if circumstances necessitated, you could be two hundred yards farther away than the rifle even, with a length of fishing line tied to its trigger.

So you see that the tricks you could play upon an enemy are many. And, strange though it may seem, with all the greater success in a locality which they firmly believe is their own.

If the enemy have taken a certain area of country and used a road over it many times, you can catch them on that road more easily than on a strange road over which they are cautiously feeling their way. Simply because they are quite unsuspecting. That old culvert over which they frequently travel, put a mine under it. That log they have seen so often lying by the side of the road, fill it with explosives. That rock they have grazed many a time, remove it and put a mined dummy in its place. That little house on the roadside, so convenient for cooking and camping purposes, mine it. And so on.

Meanwhile, don't forget that the enemy likes a drink. He enjoys beer and wine. Why not entertain him? Leave a half-emptied case of beer, a few bottles of wine or whisky or gin in a hastily vacated house.

Or, let him find the treasure .... Now, just how would you treat a case like that?

HMAS *Kuttabul*, sunk by the Japanese in Sydney Harbour, June 1942.

6

That "Spying Eye"

THROUGHOUT your sporting activities don't neglect the Enemy General Staff. These birds are big game and they generally nest back from the firing-line, sometimes in positions easy of access. This because they believe they are safe. But no enemy should be safe from the Australian guerrilla.

To destroy headquarters of any divisional, or even brigade, command, especially during active operations, could go a long way towards causing disaster to that division, or brigade. Headquarters are the centre from which orders are both received and issued. Sudden disorganization here could bring serious consequences to the enemy. Vital consequences, if circumstances favoured you and it was in your power to do the job properly. By this I mean wipe out the entire staff so that the alarm could not be given, and at the same time have a man in your band capable of transmitting fictitious orders to the enemy.

Under varying circumstances and conditions of warfare it sometimes happens that a staff is so isolated behind the security of their own lines that this could be done. It would take a large band, or combination of bands of guerrillas to do it properly - mop up the whole crew and waylay the staff cars, dispatch riders, messengers, supply men, and what-not who would be continually arriving. This wiping out of a divisional staff is no war dream; it has been done a surprising number of times by Russian guerrillas. They have wiped out a number of generals together with their staffs.

An attack on a headquarters staff would need to be carried out very much as other guerrilla operations. Good scout work first. Find the headquarters. This might come about through the fortune of war; much more likely through wide-awake observation. Perhaps the scout, watching a plane sees it drop a message above a distant clump of timber, or above any apparently well-sheltered, unobtrusive spot. Maybe he finds a field telephone wire and follows it along. Or notices a staff car, or a dispatch rider turn off from the track and vanish among the timber. Detection is up to the scouts. The next thing is to reconnoitre and learn how the place is protected; and the number of the men detailed for the job, and their weapons. How distant are headquarters from the nearest body of troops? Have they an established wireless? Is there a small landing-place for a messenger plane? Do cars come and go regularly? At what periods might they be expected? What number of cyclists are engaged etc. Every scrap of information will be of value.

On the information, on your numbers and arms, and on the circumstances prevailing, base your plan of attack. If the enemy are too tough a nut for you to tackle send your complete information at the double to our nearest body of regular troops. The job will then be up to them, probably with your co-operation.

When tackling any headquarters make sure to scoop,

all papers; they may be of extreme value. Carefully search all enemy officers, dispatch riders, and messengers. If any cars are damaged in the attack search them if possible, for secret pockets which may be concealed inside the car. And don't forget that a messenger may carry a message in his socks, the heel of his boot, in the hollowed out butt of his revolver or elsewhere;

If the enemy succeed in penetrating inland he will occupy any station homestead or group of farms or township in his line of march, which may not have been treated to the scorched earth policy. All such places, I earnestly hope, will early be evacuated by the women and children.

Some homesteads and a township or two may fall intact into the enemy's hands. Make of them a trap for him. His main body will push on; only garrisons will occupy captured places. Wipe out the garrisons. That should prove quite practicable.

The scout brings in detailed information. And you are helped considerably here by the fact that there is sure to be a local lad or two among you who knows every room, outhouse and shed in the station homestead, every track and horse pad. Some will know every house and backyard in the township; and every approach to it.

The homestead should prove a simple matter. The township nearly as simple. If you have not the numbers for the job, co-operate with a neighbouring guerrilla band. Most bush town. ships generally have only one road running through them. The largest building will be the pub. And if I know anything about the general run of army officers of any nation, most if not all the officers will be there.

Your scouts will find out the number of sentry groups and where placed; where vehicles and supplies are generally parked; the building used as a communication office; the approximate number of men in each house; where machine-guns are placed, and any other useful information.

As a precaution against any unheralded interruption, decide whether it would be best to string a few wires, with mines attached, across the road both coming and going just outside the town. It would be stiff luck if, while you had a nice go on, lorry loads of enemy troops came bowling into town. They'd be shooting into your backs before you knew where you were.

And this is where the fullness or otherwise of your information comes in. Is the enemy in the habit of sending out patrols from the town by night?

Is he in the habit of dispatching lorries or trucks back to his base, or out to the front line at night? If so, and you stretch a wire across the road then the fact may at any time be discovered from *within* the town. And a reception committee would be formed to receive you.

You could still guard yourself from interruption from without by stretching the wires flat across the road, ready to be pulled taut should vehicles come to enter the town. Outgoing vehicles, patrols, etc. could be allowed to pass over the wires. It would depend on whether you could spare the men necessary to stand by the wires.

Attack at night. Like shadows advancing. Surround the township and split up into little groups, each to your separate jobs. Some to slip into the back yards; others to worm their way along from both ends of the street, creeping in the shadows, or in pitch darkness, to the house each has to tackle. Those coming in from the yards creep towards doors or windows.

At the signal, or at the first cry of alarm, all spring up

and rush their objective; hurl grenades in at the windows; throw fire-bottles on roofs and verandas; then train your rifles and guns on the doors. There's nothing like grenades and fire to cause men to come tumbling out of a house and they come the quickest way - by the door. Spray the door with lead. Other men will be downing the sentries at the first cry of alarm.

Carefully planned, you should have it all your own way. Those sleepy men must come out. They will be lit up by the flames as they rush the doors. You are outside and wide-awake. Surprise, grenades, fire, smoke, bullets, all are against them.

With a little thought you will realize that it is quite possible for guerrillas to successfully attack regular troops when the latter are quartered in buildings. Make sure first, of course, that all our own folk, our women and children, have been evacuated. In this war captured women and children have been kept in buildings, and the brave soldiers have sheltered behind them. That is yet another reason why every woman and child should be evacuated immediately there is danger from a distant enemy, You cannot throw bombs into houses in which there are our own women and children.

The case outlined above is a simple case of attack which, well planned and carried out, would be a complete success, in ninety-nine cases out of a hundred. It could only be so, however, if every detail was first thought out and, if possible, rehearsed.

In addition to the points already elaborated, here are others which would need to be considered: How far away is the nearest body of troops from whom the enemy might expect quick assistance? How long would it take those troops to reach their compatriots if called upon?

That would mean a vital time factor to you. If you could not prevent the enemy from sending for that assistance, your time for the actual attack would be strictly limited by the time it would take enemy reinforcements to arrive. All hands, by prearranged signal, would have to break off. the engagement and make their getaway just before the reinforcements could arrive.

Unless you had combined with other guerrillas to set a trap up the road along which those reinforcements must come.

Another detail is this: The immediate objective of one group of your men is the enemy's communications - the house containing his signalling or message system. At the very first alarm this must be instantly rushed. If this is done in time it will prevent the enemy sending out word of his plight.

Carefully plan to immediately wipe out any machine-gun defence your scouts have reported. Men must be detailed to rush these and bayonet or grenade the crews, then turn the guns upon the enemy.

The position, numbers, methods of relief, of the sentries will have been reported by the scouts.

Men must be detailed to attend to these sentries when the alarm rings out. Others should be detailed to rush any vehicles, thus preventing escape.

The effect of the noise and the fires must also be thought out. Would the fires be seen by distant bodies of the enemy? If so, how long would it take them to "take a tumble and rush reinforcements along"?

You must have the place watched to the last moment, lest a travelling body of enemy troops rolls into the village and decides to camp there the night. If you attacked a few hours later it would be you who would receive the surprise.

In any attack, no matter how large or small, it is attention beforehand to these little details which will offer you victory. If you neglect details you will sooner or later land yourselves into disaster.

In defence it is the same thing. Details count-even before the actual defence. For instance: You might operate from a hide-out, or frequently use some particular position in a local area. It is a well-hidden, well camouflaged, position. You feel perfectly safe. Are you - from the eye of the aerial camera?

Even if the enemy do find your position it is such that you feel confident you can put up a winning flight. Can you from attack from the air?

Your position may be absolutely concealed from the ground; but if it is not so from the air then planes will seek to blast you out of it without giving you a chance to fight their infantry. That is where detail again comes in.

If, in the heart of thick timber you have cut down trees for a camp, the observer sees it, let alone the eye of the camera. If you have fashioned a handy little bridge across the creek and it is not shielded by branches, then observer or camera sees it.

If you have dug a little trench to hold refuse or any description, the camera sees it - unless the trench be well under branchy trees.

And the camera picks out the track made by your feet and the cooks when you go to the creek for water. And the path that's made where you regularly run in the horses. Even if you have dug a trench, or a line of gun-pits, around a hill and camouflaged them perfectly, from the air two things could put you away. The soil and the track. If you had not hidden the freshly displaced soil it would show up in an aerial photograph. It is no use carefully spreading it out, it would still be visible. If you walked from the camp to the trench, always the same way, the photograph would show the track. And a suspicious bomber would do the rest.

Even by walking through grass it would be possible to put the show away. For instance. Say that half a dozen of you were in the habit of sniping at the enemy from a possy that defied their detection. You walked to and from this possy across a grassy paddock, so as to leave no tracks. Or, say that half a dozen or so of you walked behind one another across a grassy paddock to your hide-out, and in the morning a reconnaissance plane came over, trying its hardest to find where the snipers came from, or to find your hide-out.

They would photograph the country in sections. And across one grassy paddock would be a silvery trail.

They would only have to follow it up to locate your sniping possy, or the hide-out, as the case might be.

That silver trail will show up on the photographic plate because of shadow. Where you had trodden upon the grass the blades had been bent down, even though lightly. This displaced shadow. That is, the bent stalks no longer threw the same amount of shadow as the stalks beside them. And the result is a silver trail across the photographic plate which, if overlooked, might put you away.

And now, as to vehicles, for you will be using these. You might cut across country with a truck and think that as the country was heavily grassed you'd leave no tracks. Wrong. Those tracks would show up in an aerial photograph as plainly as if they were deep brown ruts.

Across on the other side we used to build redoubts, and encircle them with barbed wire. If you built a redoubt, or any defensive position, and did it so well and camouflaged it so naturally "into" the country that it could not be seen from the air, could it then be seen?

Possible - in a short time - by the wire.

If you'd interfered with that soil in any way (perhaps dropped a little loam when you were carrying away the refuse from the trench) then the grass would grow a bit faster. And you can be betrayed in another perfectly natural way. If your position was in a paddock, and stock were still in it, they would continue to keep the grass down to normal level. But they could not get at the grass under the barbed wire. That grass would grow longer, and would show up quite black upon the photographic plate. There would be the distinct shape of 'Your barbed wire entanglement-and the secret of the position would be put away.

Whenever you construct a more or less defensive position, or camp, or trap, or ambush, think very carefully of that camera eye. It is marvellous what that wretched thing can find out. You must arrange your work so that the job fits entirely into its surroundings. It must appear perfectly natural from the air. Even if it is well camouflaged, if the plate shows some queer outline or shadow that it not perfectly natural, the show is put away. Not only will the enemy be warned, but he will come with the bombers. And all because of lack of imagination; lack of attention to detail.

Here is a little "extra" detail. Camouflage your work from the moment you start it, not afterwards.

It is useless camouflaging a position if an enemy reconnaissance plane flies overhead while you are building it. The plate will detect the commencement of the the work and the position is betrayed.

You would not be such fools as to erect tents in well-sheltered bush, and believe that the trees would hide them. They won't. The white, or light colour of the tents would plainly show up among the trees. So - naturally camouflage them. Paint them dull green, say, and shelter them under actual foliage as much as possible. Then camouflage their *form;* their shape.

And don't forget that unless your possy is exceptionally thickly timbered, any pads you make will be visible to that camera eye, even though the foot-track is down there among the trees.

Finally, you would not, of course, leave any saddlery, cook's billy-cans, etc. out where they would be visible to that aerial "spying eye".

Japanese submarine shelled Sydney's eastern suburbs on June 7 1942.

7

"'Ware Grenade!"

HAND grenades will be one of your most effective weapons. Although they are simple weapons you should learn all you can about them before handling them. Otherwise you are very liable to kill your mates as well as yourself.

The hand grenade is a short-range weapon which, when it explodes, hurls fragments all around; hurls them farther than you have hurled the grenade. Hence, if your mates and you are not under cover, you are liable to fall casualties to your own grenade, as well as the enemy.

If you throw a grenade from a trench you and your mates are under cover. If out in the open, you dare not throw unless you and your mates are under natural cover of some sort. And if, out into the the open, you threw grenades to lob down *into* a gully, you would be safe. Lie flat after the throw; you can throw kneeling or lying down.

With a little practice a man will fairly accurately throw an ordinary grenade to a distance of twenty five yards. A well-trained man can lob it up to thirty-five yards. So that the distance to which you can throw the grenade is very short.

But the *bursting* range of the grenade is greater than the distance you throw it. That is the big danger; the grenade on exploding can be deadly up to as far as a hundred yards. If you have only thrown that grenade twenty yards then, if not under cover, you will be within its field of fire.

Remember that fact, for it will impress upon you the necessity of learning all about grenades and how to handle them before you start monkeying with the dangerous things.

A grenade can be thrown from any position, so long as you are all under or behind cover. And it *must* be cover. Fragments from a bursting grenade fly at any old angle at all; they whiz in every direction and into the most unexpected places. So that if your foot, or hand, or eye, or nose is out of cover, it may stop a fragment.

Imagine the danger then to a dozen of the enemy, when your grenade lobs right amongst them.

Another grim fact which makes the grenade a terrible weapon is that it can kill the enemy even when they are behind cover. Imagine an enemy sheltering behind, say, a brick wall. You cannot see him; nor can you hit him with rifle or machine-gun bullet; nor would shrapnel get him.

But the grenade can. Just lob it over the wall behind him and he's dead meat.

But remember, this fact can be turned upon yourself, for the range of his grenade is the same range as yours. Hence -surprise. Get in your grenades first. If you made a mess of it and it came to a straight-out stoush between you and the enemy, it would depend on coolness, on observation, on accurate throwing, and on your initiative in quickly seeking new cover, or efficiently operating from behind perfect cover.

Look to your cover then if you intend to use grenades and mean to make a fight of it. Good cover is not nearly so essential if you plan merely to hurl grenades amongst an enemy then make a getaway in the confusion. But cover means *everything* if you intend to make a fight of it. Each man should see then that his cover is not only in front, but on each side, behind, and overhead if possible. If you are preparing an ambush you have a chance to arrange this beforehand. Each man will know just where he is going to lob his grenades. Let him imagine then the enemy lobbing grenades back at *him*. Visualize the positions where those grenades could lob before and all around and behind you, and then seek or make cover accordingly. In a straight-out grenade fight such foresight would not only save you numerous casualties, but would win you the fight.

If you happened to be defending or attacking a township held by the enemy, grenades, and your local knowledge, would be weapons which should beat the enemy every time. They are terrible weapons in house to house, or street fighting. You could constantly throw your grenades from behind walls, covers, practically anywhere at all, then skip to another possy and not be in the vacated one when enemy grenades replied.

Grenades, and grenade fighting sounds simple, and is often written about in a way that gives the impression there's nothing to it. Don't believe it. There's a lot of "science" in it. And that "science" must be adapted in different ways, often to altering conditions when a fight is actually on, all depending on conditions at the time, on the locality, the weapons of the combatants, and the wits of the officers commanding. Believe me, a good grenade leader is worth far more than his weight in gold to your band, he is worth many a success and a lot of your lives.

I have just said there was "science" in grenade fighting.

Let me illustrate: Imagine that your guerrilla band has attacked and successfully surprised an isolated enemy position. All has gone well. The enemy have been disastrously surprised; half were wiped out in the first attack. But the other half have rushed to a strong post from which they are hurling grenades in a desperate attempt to keep the unseen foe at bay.

To rush that post now means the loss of a lot of men under grenade, machine-gun, and rifle fire. You would spoil what promises to be a good job.

Meanwhile, you are under cover close around them, right in their own camp. They cannot see you; do not know your number. And time is with you. But what is to be done? Every few minutes, in a "delayed-time" action like that, casualties must inevitably occur. Your leader chooses a commanding position some distance away; it does not matter how far away so long as he can see you to signal and watch the strong post through glasses.

He soon gets an idea where the enemy's strongest bomb positions are. These are the danger. He can see also the ground close around the post, and the places here and there where a few of you could crawl, under cover, to within grenade range.

He sends you detailed word. Half a dozen of your best bombers crawl forward, get under the cover allotted them which is under the eye of the leader. They look about and before throwing a bomb locate another possy which the leader has seen, either behind them which will be a safety cover, or preferably a cover farther around the enemy but within bomb throw. The enemy's grenade throwing has died down; but they are at the ready. Your bombers get to their cover. The leader signals them the range, say thirty yards.

This signalling is simply arranged beforehand. The leader's attention is focused through the glasses upon the enemy, watching for every enemy move, and every suggestion of a move from the enemy strong post. To enable him to keep his attention on the strong post and 'Your bombers he will station a man behind him. He says to the man "Thirty yards".

This man signals the bombers - the grenadiers to be exact.

The signal may be an arm raised above the head signifying five yards. Raised six times the signal means thirty yards. Any simple system.

The bomber then throws. His grenade falls three feet short.

The leader instantly snaps "Three feet short!" (Or short, too far right, or left as the case may be.) And the signaller passes the message on.

The bomber then throws again, putting the extra effort into it to make the bomb lob five feet farther, or so many yards to left or right, as the case may be. If the bomber succeeds with this second grenade, it means that a grenade explodes right within some portion of the enemy strong post. Both casualties and demoralization must result.

The enemy reply. But as they have not seen your men their fire is wild and produces no effect. Your bombers may throw half a dozen grenades or more before some scared beauty in the strong post gets an inkling of where the grenades are coming from.

Your leader through the glasses spots this, and the obvious retaliation coming. "You're spotted!" he snaps, or any warning agreed upon. The signaller behind instantly signals "You're spotted!" and the bombers duck back to that cover beyond range, or to the second cover which is within range.

If the leader had not noticed preparations for retaliation, he would notice the preliminary grenade bursting a few yards ahead of your bombers. Then he would signal, and they would duck out of it - lively.

The bombers are now in their second position, still unhurt, still unobserved. The manoeuvre is repeated. The leader signals them the enemy's activities, the range. and results, and thus they carry on. They lob their own grenades right in amongst the enemy while dodging his. When grenades begin to fall into any section of their strong post the enemy throw their grenades all over the place, expecting an attack with the bayonet. Meanwhile the rest of the band are under cover some distance away, ready with the rifle to open fire should the enemy in desperation hop out and try and locate their unseen foe.

By such a method that strong post could be bombed out of existence, with possibly no loss at all to our side.

So, there is "science" in it. As always, if you do not understand your tools (weapons), what they can do and how to use them, you make a poor job of it.

In this little illustration, the annihilation of a small strong post, make an exercise of your imagination and talk over the job with your mates. Increase the strength of the enemy; vary the positions held, lay of the ground, etc. Then work out the ways by which our bombers could still defeat the enemy with probably but little loss to themselves. You will find it a fascinating war game which will save you lives when the whips begin to crack.

You will see how two relays of bombers could be used, one merely to distract the enemy's attention while the other team crept well within range. You may see a chance whereby a shower of grenades lobbing into one strategical corner, or point, of the strong post would kill the enemy there. A rush by a few of you into the smoke and confusion would put you fair into that area of the strong post. From within the enemy you could then hurl bombs into their very midst while the rest of the band, outside, by concentrated fire threatened other portions of the enemy's position.

You may see chances whereby the leader, from his superior vantage point, could place a few snipers, and thus add to the losses of the enemy, and further undermine their morale. You begin to see how you could utilize smoke-bombs for cover and so on. And to realize that what at first apparently appears a suicidal position to attack can really be taken fairly easily by what I have loosely termed "science". Which simply means that if you understand grenades and how to use them, if you have developed a quick and simple system of signalling, if you work perfectly in teamwork you will put it all over the other fellow every time.

Not only can a grenade be thrown to lob behind a wall, behind an enemy tank, behind anything against which the enemy may be sheltering, it can be thrown up to lob in a travelling truck filled with troops, or into the open turret hatch of a passing tank, down into a gully or dug-out, or up into the second story window of a house. It can be thrown practically anywhere, ruled only by the cover of the enemy, by the efficiency of the thrower, and the range at which he can throw. But accuracy in throwing is a bit more important than long-distance throwing. One bullet has the power of killing one man. I have seen two men killed by the one bullet on several occasions, and once saw three men killed by the one bullet.

Such instances, however, are only the one chance in 100,000. One grenade can quite easily kill seven men and wound a dozen others. Given the two weapons, surprise and grenades efficiently handled, you can tackle, under practically any circumstances, an enemy considerably superior in numbers to your own.

The grenade is thrown at the "over arm". That is as a cricket ball is thrown, only the bomb is thrown high into the air, to give' it range and aim. Throw it with a natural body swing; learn the throw which suits your own physical body. Practise with a dummy bomb, otherwise a jam-tin, holding, say, about a pound and a half of sand. Make a three-foot circle on the ground twenty yards ahead of you, then try to lob your bomb into that ring. Practice makes perfect. Very soon you will throw accurately, and probably increase your range. Then get behind a rock, or down into a gully and try to throw your bomb "up and over" so that it lobs behind an imaginary enemy out of sight. Then get your team together and practise bomb throwing in dummy action. You will be surprised to see how many of your own men would be hit by your own bombs, let alone by the enemy's. Each man must learn from every lesson. Each automatically seeks cover, throws accurately, and sees or knows that his comrades are under cover when he does throw.

Only by practice can your guerrilla band successfully learn how to handle grenades. Once you do know you will have increased your fighting efficiency hundreds per cent. And save many casualties among yourselves. For not only will you have learned how to use grenades, you also will have learned the danger of enemy grenades, and points in dodging or overcoming them.

Grenade fighting is a form of warfare in which leadership, surprise, efficiency, initiative, quickness and wide-awake wits win every time. Because a grenade that may lob fair amongst you could do disastrous damage, when you are practising and. discussing grenade actions, try to think out some simple means of defeating such a grenade. It should be quite possible to devise some simple means. If anything has been done about the matter in this present war, I so far haven't heard a whisper of it. And yet, grenade-fighting is still the most desperate form of close-range fighting on every front. Hence, think before on how to save your lives should a grenade lob amongst you.

My practical experience of grenade-fighting was gained in the last war. But the grenade is just the same in this war; there are more types of grenades; that is all. Their explosive power, use, and radius is just the same. Half-way through the last war we were equipped with Mills bombs. And according to military experts, the Mills bomb is still unsurpassed by the grenades of any nation to-day. So that the numerous factors which applied in grenade-fighting during the last war, apply today.

There's one interesting little point I'd like to mention. When my old regiment landed on Gallipoli we took with us a few old-type cricket ball grenades. These soon ran out. And as the Turks were blasting hell out of us with modern grenades we had to make grenades of some sort, so we made the good old jam-tin bomb. At the same time we had a few cases of very nice, metallic, beautifully finished Japanese bombs - a present, so far as we ever learned, from the Japanese War Office.

While they lasted, those bombs were the terror of the Turks. That was May 1915. They beat all our bombs into a cocked hat-until the Mills came along.

We took, gave and dodged grenades under all manner of conditions, just as you will have to do. In open-country warfare, rocky country, orchards, hedges, trenches, dug-outs, desert, wherever and under whatever conditions fighting was going on, by day and night, we instinctively became used to various warnings, or no warning at all except the soft plonk of the bomb falling near by. Some grenades, though, had their own characteristic when coming through the air, a sighing, hissing, or whistling. Others gave no sound whatever, they just came.

Night is the deadly time for grenade-fighting because under cover of night both sides can come much closer. At night-time, when grenade-fighting might be expected, our eyes were continually roaming at an upward angle over the enemy's position. We weren't thinking of his position, but because at a certain angle an object coming through the air appears more visible than otherwise. We'd catch a fleeting shadow in the air as the grenade came hurtling towards us. We did not see them all by any means, but everyone we saw raised a warning cry. It all depended on our adaptability of eyesight, and the night light .and sky as to whether we saw them or not. The open sky, of course, was the background across which the fleeting shadow of the bomb had to come. A floating cloud would often hide it.

Grenades never come straight at you; they are thrown up into the air to fall in an arc, as it were. Otherwise you couldn't see them. In the same way, under favourable conditions, you can see an artillery shell travelling through the air. Many a time in broad daylight we've watched the flight of a shell. On a dark night, any bomb of a "fire-fuse" type was often betrayed by a few tiny sparks as it came hurtling, generally over and over, through the air.

Some grenades would burst instantaneously as they hit the ground. There was no hope against them, except by the millionth chance. The majority didn't explode until a couple of seconds after they lobbed amongst us. You can do a lot in two seconds-a-if you're keyed up. In three seconds 'You can be a "mile" away. Our minds made our bodies act with the warning cry while the bomb was still in the air. If there was shelter, we'd leap into it. If not then a leap away, and another and another leap, and flat down into any tiniest depression in the ground, or behind anything at all, with a greatcoat flung over us. If there was no getaway then every hand grasped and lifted overcoat or sandbag and when the grenade fell a dozen sandbags and greatcoats smothered it.

They were primitive precautions that we took with lightning swiftness. But there's many a man alive today who would not be alive had we not been ready.

It may seem impossible that a grenade could land in a narrow trench filled with men and that any within radius could survive. The explosion alone in that confined space should kill all hands within the limits of the explosion. But I've seen men survive again and again. In Lone Pine, for instance, which for months saw perhaps the most terrible grenade-fighting in the world's history, the Turkish and Australian trenches were only a few yards apart. Grenades of all descriptions were coming and going day and night. Yet a couple of sandbags, an overcoat, instantly thrown on a grenade followed by a desperate crouch trying to push the side of the trench away, saved many a man. A grenade on which a greatcoat is thrown may kill five men and wound others, or it may hardly wound a man, it depends on how the coat is thrown, on the loose folds of the coat, and on the position of ground and grenade below it. The thick loose folds of the coat, and on the position of ground and grenade below it. The thick loose folds of the coat slow up and block many pellets from flying from the grenade. A long, thick overcoat will, to some degree, even confine the explosion - as with sandbags. There is the terrific explosion, of course, and inevitably some men are knocked but others are saved who would not have been.

Even the "smother up" may save a man. Hunch up, crouch like a crouching monkey into the smallest thing on earth. Sink your head deep into your chest with close shut eyes, one hand and arm protecting face and chest, the other hand and arm protecting stomach and what you think most valuable. Then turn your back and pray. Such precautions may sound ridiculous. It's not. It is the gambler's chance. Twenty bomb fragments may miss you by fractions of inches. Whereas if you'd been standing they would have got you. Besides, miracles can occur.

What I've written is just a lead. In a spare moment or two think out some simple precaution which will save you and your mates from grenade explosions.

8

More about Grenades
and Bombs - The Trench Mortar

GRENADES are of many different types and shapes. Some have handles attached, which give greater throwing range. Sawn-off lengths of broomstick make excellent handles; the Turks taught us that with their Broomstick bombs; Russians and Germans are using them to-day, Some grenades explode by fuse; others by contact; most by the release of a pin and firing lever which allows the grenade to explode in a certain number of seconds, from five to seven. The outside casing of some grenades are deeply grooved, dividing the outer casing into squares or diamonds. The object of this is to ensure that the grenade, on explosion, will shatter into as many sizeable fragments as possible.

A home-made bomb can easily be made out of say, a foot-length of inch iron piping. Plug one end; drop in a few broken iron fragments, then a couple of plugs of gelignite with cap and fuse attached; plug the other end. Attach a two-foot length of broomstick handle to one end if you wish; you can then throw it farther. The broomstick acts as a tail which steadies the flight of the grenade. This makes a nasty bomb, easily thrown. But, as the casing is not grooved, it breaks into far fewer fragments. Grooved, it would be far more formidable.

You can do this with an ordinary cutter. At every half inch, cut a groove around the pipe nearly as deep as the metal. This makes the grenade "anti-personal".

The handy grenade generally weighs a pound and a half. Heavier grenades require; of course, more gelignite, dynamite, gun-cotton, or whatever high explosive is used according to their size.

Various nations favour their own type of bombs and grenades. Perhaps the British Mills grenade (or bomb) is the most efficient in the world.

Since the Spanish Civil War great use has ben made of "fire-bombs", "fire-bottles", "Molotov-cocktails", "smoke¬bottles", "phosphorus-bottles", "battle-grenades", etc. These came overnight into action when the people of the invaded countries got their blood up. They were used, for instance, by the guerrillas and people's armies of China, Spain, Russia, and the Balkan States. And they were made and used because bottles were practically everywhere, while petrol was always to hand. The bottle is the grenade container, the petrol is the explosive. In the case of all these bottle grenades, their main damage, of course, is done b-y fire. Any "bottle" can be used, from a lady's cream jar to a huge jar for storing oil or wine. Those too heavy to throw are used as "fire mines". Sauce, beer, gin, whisky bottles make good containers; a full beer bottle (say) can be thrown about twenty yards; a half-bottle thirty yards. Before using the bottles make sure that the original contents have been dissipated.

When the bottle breaks upon the target the inflammable liquid catches fire and burns fiercely.

These weapons, skilfully used at close quarters, could

do considerable damage to anything inflammable. Supply dumps, lorries, grounded aeroplanes, houses, for instance. A score of them thrown on to a tank that had fallen into a trap would make the air inside decidedly hot for the unlucky crew. The early forms of fire-bottle all contained simple petrol.

Now, other mixtures, such as creosote are added, to give to the petrol a still fiercer inflammability, and in some cases a stickiness, so that when the bottle crashes the flying liquid clings to the first surface it touches and fiercely burns. Soft pitch, tar, tallow, sulphur are often used in Molotovs. The petrol used is generally fifty per cent. If the bottle is scored with a glass cutter it breaks much more easily. A pretty efficient mixture for a Molotov is one-part petrol, one-part paraffin, two-quarts tar. Mix and bottle.

Just like that.

The original fire-bottles were fired by a rag soaked in petrol which entered the bottle through the cork, or merely by a petrol-soaked rag tied around the bottle. They were game men who handled them, for they often exploded before time. Don't you try such a method of exploding a fire-bottle. A better way is by a fuse inserted through the cork.

The bottles now are exploded by various, and much safer means. Generally from the bottom, or side; by means, as a rule, of time fuse or cap. Cut the fuse to five or seven inches as in the jam-tin bomb. Insert fuse end into cap. Fix this securely to bottom, or bottom side of bottle with a narrow strip of canvas, or adhesive tape. Light the fuse and throw.

Some fire-bottles are set off by ignition fuses, a match-head fuse, or various other simple means.

"Smoke-bottles" are used similarly. The bottle contains chemical compounds which, when released to the atmosphere, create a cloud of smoke that may last long enough to obscure the vision of an enemy in tank or other vehicle. This simple smoke-bottle can be used in a number of ways in very close range fighting: to give momentary cover so far as obscurity from an enemy is concerned; in street fighting; or manoeuvring for position by a sudden dash. It really is a tiny miniature of the giant smoke-screens of the battling armies; it only lasts from about half a minute to one minute, depending on the wind. Used under the correct circumstances, it would shield from observation the dash of a man or men to other cover.

And now something about trench mortars. They will be your artillery. To obtain them you take them from the enemy.

The projectile from the mortar is a bomb considerably larger than the hand grenades. Mortar bombs are of various types: high-explosive, smoke-bombs, star-bombs. Smoke-bombs, of course, are to form a smoke-screen. Most bombs are of the high explosive type. The trench-mortar bomb possesses a vastly greater range than the grenade, and a greater danger burst; that is, there is danger well within a hundred-yard radius of the bursting of the bomb. The ranges of mortars vary: some may have a range of three hundred yards; others range up to a mile.

While the ammunition lasts the mortar can be fired very quickly. In fact an expert crew can keep a stream of bombs "pumping" from the mortar. Hence it is a very nervy job for attackers who must pass through a bomb screen thrown by trench mortars operated by expert crews. Very nervy, too, if you are holding a fixed position under a well-directed fire of mortar bombs. Realize then that the mortar is an extremely dangerous weapon with a fairly considerable range, and is a weapon that you can use. Take it from the enemy and use it against him.

The mortar is a very simple weapon: An iron tube mounted on a tripod. Modern mortars are on a bipod. You aim the tube up at an angle towards the enemy; drop the bomb down the barrel and immediately it is on its way.

That is the simplicity of the thing. As in other things, attention to detail gives greater understanding and efficiency. We'll take the bomb first. To the base of the bomb a propelling cartridge is attached, a rather similar arrangement to your rifle cartridge; that holds the propellant charge which fires the shot or bullet. Now, your rifle cartridge has a cap, which explodes the charge when the striker (moved by the trigger) hits it. The cartridge on the base of the mortar bomb has a similar cap. With types of modern mortars a striker is also attached.

The barrel of the trench mortar is an iron tube with its vase "closed in". This base in modern mortars is screwed into a breech piece, which in turn is connected to the base plate. And in the centre of the breech piece is an iron stud. You slide the bomb down the barrel base first. When the cap and striker on its base hit the stud the cartridge charge explodes and the bomb rushes up into the air.

That is the business in a nutshell. There is no need to go into all details. The mortar is a weapon everyone of you can pick up and quickly learn when you grab one. Still, a few points may not come amiss. For instance: Don't gape over the muzzle once you have let go the bomb. If you do 'You'll be minus a face. Again, when your hand allows the bomb to slide away, withdraw your fingers like greased lightning.

To give a bomb greater range, an extra cartridge can be fitted to the bomb. Most bombs are fitted with a tail vane (like an aerial bomb) to give them "steadiness", "direction", etc., in the air.

The method whereby the bomb itself is exploded

varies with the types of mortars and bombs, and to some extent with the nations which use them. This will not be of great importance to you. You may be equipped with a mortar. In that happy event details would be explained. Otherwise you will have to take your mortars from the enemy. Then it won't matter by what method they explode, except that by observation and experiment you must find out before you can use their weapons efficiently,

Some bombs are exploded by fuse, others by a striking head on the nose of the bomb which when it strikes the ground jars a striking pin and cap which explodes the charge within.

As a safeguard against accidents, such as premature fire, most modern mortar bombs are protected by a simple mechanized device which holds back the striking pin until the bomb crashes on the ground. The knowledge of this safeguard should make you wary. For instance if you captured an enemy mortar and were in a hurry to turn it against the enemy make sure that the flight of the bomb will be clear of nearby trees. For if it strikes a branch it is liable to explode above you.

When you capture your first mortar there probably will be one or two little points to unravel before you get the bombs into action. For instance, there are types of mortars fitted with a safety-catch on the nose. This catch must be loosened, before the bomb can explode. Others may be dressed in a waterproof cover around the tail; or the cover may be contained in the bomb carrier. If so it only has to be untied. Otherwise the cover must be taken off the bomb. The mortar itself may have been in action a good while. See then that the breech piece is tightly screwed to the barrel end; sometimes this breech piece may be worn a bit, and the explosions will cause it to gradually unscrew. A nasty accident could possibly result. The striker stud should be clean and there should be neither dampness nor oil in the barrel. The tripod (generally the bipod) and general mounting must be firmly placed.

The actual aiming along a mortar sight is similar to aiming a rifle. But the sight and its supporting gadgets may appear a trifle complicated. Once shown, however, you will quickly learn how to use them. The mortar barrel itself points at an angle into the air; so you do not aim along it as you would along a rifle-barrel. Sight and elevation, and levelling, are manipulated by a few screws. The mortars (the bombs), are generally packed in cardboard cylinders.

It requires a team of six or seven men to efficiently handle a mortar. They can do great damage with it with a little practice. For guerrilla use it has two drawbacks: keeping up the supply of ammunition, and transport. You would be dependent on the enemy for your ammunition, so that would be his concern. Transport is a different matter. If you have neither wheeled nor horse transport, the mortar crew must carry the machine. This slows them down if you must make a quick getaway.

The mortar can very quickly be assembled and unassembled. This leaves to be carried the base plate, barrel, mounting and bombs. The sight is detachable and easily carried. A regular army crew would also carry a spare parts bag. For ease in carrying the parts a harness webbing is used.

So that to transport your mortar by foot, one man

must carry base plate and dial sight, another the barrel and a few spare gadgets, a third the mounting. Yet others of you would have to carry the bombs. A man who carried six bombs (say, for a three inch mortar) would. be carrying 60lb. weight. His cobbers would be carrying approximately 45lb. each. You would overcome the bomb carrying difficulty by each man in the whole crowd carrying a bomb.

Fire your mortar from below cover. If you don't you'll soon find machine-guns, mortars, and artillery concentrated upon you. This necessity (of keeping below) may of course affect your aiming the weapon, your sighter may not be able to see the enemy target. Pick an object that you know is near the enemy. If you believe the enemy is concealed fifty yards to the left of that object, or two hundred yards beyond it, manipulate your barrel to lob the bomb accordingly. Watch closely where your bomb bursts, for that will be your sighting shot. Make your corrections from that shot, lengthen or shorten range to right or left as the case may be. The main thing is to get your direction; that is, the direct line between you and the enemy.

If the enemy is invisible, but you reckon his possy is one hundred yards to the left of that big old gum away out in front, well, that gum is a standing mark which gives you direction. Your aim will be one hundred yards left of the gum. You must now lob your bombs on the enemy's position by sighting shots, that is ranging. And get your bombs lobbing there as quickly as possible, not only to annoy the enemy but to prevent him from annoying you. Quickness and skill in lobbing your bombs right in amongst the enemy is imperative. Otherwise he has so much more time to train his weapons upon you, and judge your direction from the bursts of your wasted bombs.

A mortar can be used at night without putting the show away, for it shows very little "muzzle flash". If properly concealed, the harassed enemy would try in vain to locate from whence the bombs were coming.

The mortar is not too good against a moving target; that is, against a target which is not actually dawdling, because the bomb is a long time in the air. It may take perhaps twenty-five seconds to reach the target. This would not matter a great deal if you had to range to a road for instance. You could still play the dickens with an enemy convoy. If the target was a solitary vehicle coming along, your object would be to estimate its speed. Say your bomb takes twenty-five seconds to reach a certain spot on the road. Then estimate what distance the car is travelling in twenty-five seconds. When it is twenty-five seconds away from that spot on the road you fire the mortar.

The mortar is a very handy weapon for keeping the enemy's heads down while the boys work around to surprise him in flank. But you must have experience here or you will smash your own men, just as you would if you threw a grenade. The bursting radius of a mortar bomb being considerably greater than that of a grenade, no man not under cover is safe from a mortar bomb if it bursts one hundred yards from him; the extreme danger zone is greater still. Therefore if you were using a mortar to cover your cobbers in an advance you would cease fire when your mates came to within not less than two hundred yards of the bomb bursts.

Mortars of all types are being increasingly used as a short range artillery. You are sure to come in contact with them. Seize all you can and immediately turn them against the enemy.

9
Minefield on Land - Grenade Fighting

IN modern war, with each swiftly passing month land mines are being used in ever widening ways with fast increasing rapidity. To-day some sectors of oversea fighting fronts are mined almost with the intensity of minefields at sea.

Seems a shock, doesn't it? We know that our harbours are mined. It means almost certain disaster for an enemy ship to approach our harbours; as it does for our ships to approach theirs. We are familiar, too, with the fact that in many areas on the seven seas, many miles from shore, extensive minefields are laid, minefields which may stretch for hundreds of miles. Also channels and various other strategical waterways are mined.

But we do not quite realize that armies are similarly mining their positions. Still, that is so. Many hundreds of mines may be laid upon a road; many thousands around a fortified position and along the approaches thereto. It is a fast developing modern phase of land warfare, and we must wake up to it. Forewarned is not only forearmed; it means also that we can lay counter mines against the enemy. So far as the guerrilla is concerned it means that, our scouts having spotted the placing of enemy minefields, in the dead of night we can dig up those mines and use them against the enemy. Plant them in the roads *he* is using. Plant them in localities over which *his* troops are passing. Plant them on airfields where a descending plane will sooner or later set them off.

Mines and minefields, unknown to us, may prove of great advantage to an enemy; but once we know about them, they will prove of far greater advantage to us. The enemy has to import his mines from great distances. Let him. We will steal his mines and use them against him.

Mines, of course, in a quaint old form have been used more or less from the earliest days of gunpowder. In the last two years of war they have developed into varied and infernal machines of war.

Mines are of different types and sizes. There is the mine which, planted in the road, explodes when a heavy vehicle passes over it; the mine which only goes off when a string or wire is touched - it "pulls the trigger"; the mine that goes off instantaneously, or by time fuse; and the mine which explodes by pressing a distant button. There are minefields, also, which go off by sections.

By night you might trip against a hidden wire, and a mine explodes amongst your comrades. Or, your stealthy approach is detected. A hidden enemy presses a button and mines go off all around you.

To plant a series of mines along a road is now an art.

The least sign of displacement of gravel, soil, or surface and the enemy convoy pulls up, alive with suspicion. They reconnoitre (which is a chance for your ambush). Your scouts therefore must very closely watch for evidence of enemy mine- laying. If mines are being laid the scouts should memorize as much as possible every position, so that the mines can be removed at night and planted elsewhere - in the enemy's path.

The scouts must watch also for distinctive marks. Not so much on the mines themselves as signs to the enemy that they are approaching one of their own minefields. It would be disastrous for either side to lay a hidden minefield, only to have one of their own convoys ride straight into it.

Minefields at sea are charted. The danger areas are known to every captain. Land minefields can be mapped, of course, but it is not nearly so easy to let leaders of all troops know of them. A captain, his crew and ship are all together, a warning to the captain of each ship is all that is necessary. Not so on land. There are army corps, divisions, brigades, battalions, regiments, companies, convoys, and supplies, operating all over the place.

To guard against this danger and difficulty some armies have adopted signs that warn their men when they are approaching their own buried minefields. And signs which point out to them the channels through which they can safely pass. It is the bushman's idea of blazing a track for others to follow. The job of your scouts, then, is to watch out for these distinctive signs or marks. They must also find out those marks which indicate danger and safety respectively.

The rest is simple. In the night-time you come out of your job and transfer the signs: put the safety post, or whatever manner of sign it may be, where the danger sign was, and the danger sign where the safety was. Thus, without firing a shot you may blow up an enemy convoy, or a battalion of their men. And at their own expense.

We can approximately describe mines as light, medium, and heavy, according to the varied jobs they are specialized to do. Whether for the blowing up of troops, or mechanized vehicles. They are metal containers in which are packed high explosives such as dynamite. Some mines will explode under pressure from a foot, under touch of the hand, under pressure from a vehicle. Yet again a mine may be exploded by means of an instantaneous or a time fuse. Or it may be exploded by wire connected to a battery a hundred yards or a mile and more away. It may be mechanically set to explode by contact with a trip wire, or by the force of a pull (such as when a man lifts a booby trap); it may be exploded by the opening of a door, the closing of a window. There are very many ways. A mine may be designed to blow up the man who touches it, or a tank, or even a fortification. But a different type of mine is needed for the last. Modern mines are rapidly being developed to blow up practically anything.

Mines may be set upon or around a road, path, bridge, or railway line. They may be set in farms, houses, complete towns. Circles of them may be laid around an entrenched position, or in a semi-circle as the case may be, thus protecting the troops within, just as sea mines are set to protect ships in a harbour.

Thus, when troops advance to the attack, they fall into a minefield which erupts in a series of explosions around and among them. The troops set off many of the mines by their own contact; others are set off by wire from the distant fortifications.

Mines when set are, as a rule, carefully buried. Often,

however, they are not. An innocent looking bucket, a plate, a cup may actually be a mine, or booby trap. So that this aspect of land-mine warfare allows plenty of opportunities for you to use your imagination. For instance, if you filled a narrow iron pipe with high explosive and set it to explode on contact, then tied portion of a dead branch of a tree to either end and covered the tube with dried bark so that it all looked like a withered limb fallen on the road from a tree above, when the enemy tank came along neither driver nor officer would suspect it to be other than what it looked like. You could easily get an actual hollow branch and shove the mine up into it. Then lay the dead fragment of branch across the track just below the very tree (or apparently so) from which the branch had fallen to the road.

Some bush roads have stone culverts or gutterings or crossings. You could easily imitate one of these cut stones with tin or bark or even cardboard with a bit of dust scraped off if need be from the actual stone beside it. Put this false stone carefully in place. No one would ever suspect it was a mine- until it goes up.

There are countless ways. Think them out for yourself; by doing so you will "take a tumble" to ways in which the enemy may seek to catch *you*.

Buried mines are carefully set just underground, in positions where the O.C. thinks it most likely the enemy will pass by, or where he is most likely to attack. All traces of disturbed earth are of course removed, everything must be left exactly as it was before, in fact the surface just there must, if possible, be made to look actually more "natural" than it was before.

Dingo trappers would be interested here. Those boys lay their mines (traps and poison baits) in such a way that the most suspicious, cunning, sniffing, sharp-eyed dingo will never know a trap is there until--

Mines will not be used as extensively in Australia as on other fronts for the simple reason that an enemy's sea transport will be taxed to the uttermost. But he will certainly use mines. Capture as many as you possibly can from him, and use them against him.

To return to the hand grenade, discussed in a previous chapter. This will be one of your best weapons. Unless you are a member of the V.D.C. you will have to make grenades for yourselves until you are able to capture supplies from the enemy.

The jam-tin bomb is very efficient, and extremely simple to make. You must have, however, gelignite or dynamite fuse and caps.

These should be easily procurable in a number of districts, for they are used extensively in mining, quarrying, road-making, building, harbour works, and other activities.

We made these bombs on Gallipoli and used them for months against the Turks, until at long last we got supplies of Mills bombs. Even though the Turks used manufactured bombs and grenades against us our jam-tin bombs kept them at bay. So their efficiency is well proved.

All we wanted was our empty jam or milk-tins, a lump of old iron or some broken bolts, nuts, rivets, spent bullets, fragments of shell case, old nails - any old metallic fragments. We wanted also some sticks of gelignite, a coil of time fuse, a box of caps (detonators), and any fine wire we could pull off an old case.

We'd break the fragments of iron into sizes roughly from a walnut to a pea; spread a layer on the bottom of the tin; put the plug of gelignite (with cap and fuse attached) in the centre of the tin, then pack around it and above, with more pieces of broken iron. We packed all as tightly as possible, taking care not to jar the cap within the gelignite. By packing the fragments, not only did the bomb contain more pellets, but the compactness strengthened the explosive power. Never use a hammer or any other instrument to hammer the pellets compactly; you might jar the cap and blow yourself to pieces. Pack the pellets together with the thumb.

The lid of the tin had a hole in the centre, through which the fuse protruded. We'd close the lid tightly and twist wire all around the tin. Our bomb was then ready for action, Generally, we'd use one plug. of gelignite per tin; two plugs if we had plenty of gelignite. The fuse used was the ordinary time-blasting fuse, such as all miners and quarrymen use. There is an instantaneous fuse also, but naturally, to use this would mean instant death. So, even in this simple matter, understand fully what you are doing before you begin monkeying with explosives.

An instantaneous fuse is one which will set a distant explosive charge off, practically at the touch of a match. As you are far away from the explosion in that case, you are quite safe. When you light a grenade you are only five to seven inches off the explosion. A time fuse is one that is timed; that is, takes a certain time to burn a length of each inch. The general fuse we used on Gallipoli, and I have used for years in mining, is one second to one inch. Therefore, such a fuse takes five seconds to burn five inches in length. A similar fuse seven inches in length burns through: in seven seconds. These are the two lengths of fuse we generally used. The five-second just gave safe time for a wide-awake man to properly light the fuse, swing the arm and throw with judgment and aim. By that time the bomb was travelling through the air to lob in the enemy's trench, or among him. A second later it exploded. Many of us liked the seven second fuse, because the extra two seconds gave us ample confidence in which to make a safe and good throw.

If the fuse was much longer, it would fall amongst the enemy with the fuse still spitting. This might give sufficient time for a quick and brave enemy to instantly seize the bomb and hurl it back again.

So, test your fuse. Cut off a length, fire and time it. You then know the rate at which a fuse burns. Cut your bomb fuse accordingly.

An instantaneous fuse would only be used on a mine you have planted a considerable distance away. You then retire to safety. When ready to fire the mine, light the fuse and instantly up goes the mine.

It is always a good policy to test a fuse. Your fuse may be old or have been improperly cared for. It may have been left in a damp place, or atmospheric or other causes may have deteriorated it. If so, parts of that fuse would smoulder slowly, other parts burn faster. So that, with a deteriorated fuse you never could tell how slowly or how fast the lit fuse in your bomb would burn, or when that bomb would go off.

Hence, test your fuse. We cut the fuse into five-inch lengths to burn through in five seconds, and into seven-inch lengths to burn through in seven seconds.

Now, when you cut a fuse (with a sharp penknife, razor blade, or such like) cut it off "square". Not at an angle. The fuse-end should sit flat into the bottom of the cap, so that the last grains of gunpowder in the centre of the fuse will be sitting squarely upon the fulminate of mercury which explodes the cap. If you made a slanting cut, the gunpowder would be just 50 much higher up in the fuse, and in that case, possibly, would fizzle out without setting off the explosive in the base of the cap.

The cap is made of copper and shaped like the cartridge case of a .22 long rifle bullet-a bit longer, stronger and larger. Inside, down in the base of the cap, is a pinch of fulminate of mercury. When the last burning grains of gunpowder in the fuse blaze on to this the fulminate fires the cap which explodes the gelignite. And heaven help any living thing within range. Now, these caps are packed in sawdust; partly to keep them dry, partly to keep them from shock against one another should the little box be dropped. Because the caps can explode violently.

Often, there are grains of sawdust within the metallic cap itself. Shake this sawdust out. Some grains are sure to adhere within the cylinder, you can bet your sweet life, right down to and on the fulminate itself. Blow into the cap, and shake again. This very probably will shake out the last few grains of sawdust. If not, it is advisable, although not always necessary, to get these out. I've done it many a time by gently poking at them with a straw, a green straw, a green. blade of grass. If you make a hard jab at them with a hard, crisp straw you may explode the fulminate and away go your fingers.

Having shaken or blown out any remaining grains of sawdust, you insert the fuse gently down on to the base.

Now, the fuse nicely fits in, but still is a bit loose. It must be firmly gripped boy the cap otherwise when handling it you will pull it away from the fulminate.

To make it firm, burr the edges of the webbing (the end of the fuse) just a little; that is, tease out the edge of the cut fuse with your thumb-nail, then push the fuse down. It grips much firmer, but not quite firm enough; while inserting cap and fuse into the gelignite and later when handling the bomb yon still might easily pull out that fuse. Miners gently squeeze the upper rim of the cap to the fuse with a pair of pliers. Imagine you held an empty cartridge shell in your hand. You would put the pliers to the open "mouth" of the shell and gently squeeze it. I've seen men do it with their teeth; put the open end of the cap between their teeth and gently press the rim of the cap against the fuse. If the cap exploded, of course then their teeth would blow away.

If you haven't a pair of pliers the simplest way and the safest to make "all fast" is to squeeze a tiny piece of paper between cap and fuse. Just push it in with the point of your pocket-knife and the fuse is jammed fast within the cap. It can be done in fifteen seconds.

The next thing to do is to insert the cap in the gelignite. That is generally called a plug, or cartridge. It is wrapped tightly around with a thick, preservative paper. Undo or tear off one end of this and, with the small blade of your penknife bore down into the gelignite, churning out a piece; it is soft, like cheese. The piece cut out should leave a hole just large enough for you to push the cap right down into. Thus your cap, fuse, and plug are fixed. You carry on then by putting the plug into the tin, just as described. When the tin is wired up, the bomb is ready for action.

Now, here is a last little point: You may not use those bombs for some time. Consequently when you come to fire one, the tiny spiral of gunpowder in the tip of the fuse may be dusty, or may have been cut any old way at all. It cannot be expected to spark at the touch of a cigarette end or a match.

So, just before you think you'll be using those bombs, nick a tiny slit a bare fraction of an inch down the centre of the fuse tip; cut into the powder, then gently pull the cut a little apart. You thus expose the fresh powder. Into this press a little blob of gelignite about the size of three match heads; press it firmly down on to the gunpowder; make it stick.

Now, when you touch that fragment of gelignite with the lighted cigarette end it will immediately blaze and let the gunpowder on fire. Wait until the powder catches with that steady "hiss-ss-sss" (it should be almost instantly) then lift back your arm and throw.

And good luck to your aim.

These few points about explosives might help.

High explosives possess terrific shattering force but yet, with ordinary care are quite safe to handle. Few of them will "go off" if you drop or knock them, though it is inadvisable to do so. However, anything is liable to happen if you knock a H.E. that has become deteriorated. Age and exposure to weather cause deterioration in which case the explosive becomes very unstable.

Should you put a match to gunpowder you know what will happen. But you can put a match to gelignite for instance and it probably will not "go off", unless deteriorated. Hence, though care is at' all times necessary in the handling of high explosives there is no need to be frightened of them.

It is a very different matter with detonators remember. Drop these on to a hard surface and they are very likely to explode. Knock one, and "up" she may go. The detonator's job is to give the shock to the H.E. to send it off. So treat the detonator carefully lest it send you off.

Gelignite will almost certainly prove to be your best H.E. With ordinary care it is very safe. to handle, can be used on practically any demolition or wrecking job, and is far the easiest to obtain.

The fuse is a tube, as it were, holding fine black powder. The powder is protected by being wrapped around and around by strips of varnished tape, and gutta-percha as protection against dampness. A coil of safety fuse will be in colour black or blue. An instantaneous fuse will be coloured

red, or orange. Test your fuse before using. Some fuses burn a little faster than others. Cut off one foot length, light it, then time it. You then learn how fast the fuse burns per inch, or foot, etc.

The end of a coil of fuse may have been penetrated by dampness, but the remainder of the fuse may be quite serviceable. Cut a foot length from the end, throw it away. Test the freshly cut end of the fuse.

To fire a charge by electricity an "electric" detonator is needed. In it is a fuse wire from which two leads extend. To fire the charge an electric current is passed through it. A car, or even torch battery can be used.

To blow up, or cut, or fracture anything at all, you must, as far as is possible, contact the object with the explosive. For instance, say you wished to cut through an iron pipe, or girder of a bridge. You would place the explosive hard up against what you considered the weakest portion of the girder. If you wished to cut through a railway line then place the explosive tight against the rail. If you found it awkward on account of the shape of the rail or whatever the object was, then where the explosive did not touch plug the "vacancy" with clay. And cover all the explosive with clay afterwards, to "confine" the explosion as much as possible.

You do far greater work with explosives if you can introduce the explosive into the object, and use much less explosive. Just by way of explanation. Say you wanted to blow up a huge granite boulder. If you laid a pound of gelignite against the boulder, flattened it down with clay, then fired the charge you would split the boulder. But if you first bored a two feet deep hole into it with hammer and drill, then two plugs of gelignite would blow the boulder into a dozen pieces.

Hence, if you wished to bring down a tunnel, or embankment, you would bore your holes first, fill them with gelignite, tamp them down well. Otherwise the job could not be done, or else would need a tremendous amount of explosive.

Japanese midget submarine retrieved from Sydney Harbour
following their attack in February 1942.

10

Two Guerrilla Weapons - Cunning and Resource

THE Australian guerrilla will be called upon to do a great deal of night work. The enemy then will be considerably more at the mercy of your cunning, initiative, and weapons than he will by day. Night to him in this strange land will be a terror. That will be the first gain to you. He will be nervous, whereas to you night will be a confident cloak.

Night allows you to worm your way through his lines, to creep to within yards of him. Under cover of night you can walk right up to one of his tanks, to his headquarters; you can peer down into his trenches and hear the whisper of his conversation. You can lay mines under his very nose, can fire his dumps, can do many a daring thing you could never do in daylight. By night, his automatic guns and grenades lose their terrors for he cannot see you to use them. Even if by a false movement you betray yourself, he fires blindly. And you can catch him under circumstances in which, if he did open up, he might do more harm to his own men.

Your advantage by day, in that you know the country, becomes incomparably greater by night.

But you must *really* know the country. You may know it by day, but that same country may appear very different by night. Try it out. Make a night trip across bush that you know very well. You'll be surprised at how different the bush, the distances, the sky, the old familiar hills and gullies, look at night. Often they appear deceptive. And different again according to whether the night is starry or cloudy, windy or rainy.

I've been asked from various quarters to write all I know about scouting, bushcraft, and nightwork. I'm afraid I don't know a great deal, but after many years in the outer bush I feel competent to give a few hints. What little I know about night work will go into a small book on scouting that will follow this one. Meanwhile I strongly advise you to turn to the best of all teachers - experience. Get in all the night work you possibly can.

Here are a few points that may come in handy sometime. Do not over-estimate your strength. It is fatal. On the other hand do not under-estimate it; that would rob you not only of a percentage of usefulness, but of striking power. If you under-estimated your strength you would miss chances of smacking the enemy hard.

A guerrilla must be a dashing fighter. Instead of timidity, he must develop cunning allied to boldness. And then, his plans laid, he strikes hard and viciously.

A day will come when, among other things, you capture a number of enemy smoke grenades. Sooner or later you will use them. Watch the wind. First be certain of the direction in which the smoke will drift, otherwise you may blind yourself instead of the enemy. Around a tank for instance you may make a lovely smoke-screen and quite hide the tank. But if the wind were not blowing it "into" the tank and all over the tank the crew might hop out under cover of your own smoke-screen and lob grenades amongst you. So, in every instance, learn the uses of your tools before you handle them.

Are you good at boring a nice round hole in a lump of coal? It's a difficult job, but there are other ways. Put a plug of gelignite with cap attached, or a helping of dynamite into the coal; cover up all traces so that it is only a lump of coal; then place it among the coal in the bunker of an enemy locomotive so that the driver will shovel it into the engine.

Can you tap a field telephone line or a telegraph line? The enemy will use them.

The Military of different nations have devised an "unbreakable" glass for use in motorized vehicles of various sorts. Sometimes this glass is really unbreakable by a bullet, but it can be given a shattering smack from a simply doctored bullet. With the point of your penknife bore, by twisting the knife, a funnel-shaped hole into the nose of the bullet. When this hits the glass it may not break, but the shower of flattened lead blurs its transparency so that the glass is useless.

Do you understand time bombs? They are ordinary bombs with a gadget affixed by which you can set them to explode at such and such a time. They are handy for blowing up ammunition-dumps, petrol-dumps, or aeroplanes. You set the bomb to explode in half an hour, an hour, or more as the case may be; then plant it where it will do most harm. This gives you plenty of time for a getaway.

Time bombs also give you a chance to do far greater damage. You may have spied upon an enemy camp; found out where their dumps are. They will be wide apart. At night you could put a bomb in each dump - and an inefficiently guarded aerodrome. By night a crowd of you, probably, could rush that drome and do a great deal of damage. The catch might be difficulty in making a getaway, Could you all get to safety, miles back beyond the enemy's lines? If not, then such a job might much better be carried out by a very few men loaded with time bombs. A bomb planted in the "innards" of each plane, another or others among the aerial bombs. You could all be safely away before the bombs began to pop.

Do you want to stop a tank? Put a dummy mine on the road.

Do not forget that while horses don't need refuelling depots, they appreciate a feed all the same. Try and line your neddy's ribs with a few feeds of hard tack before you take him on a long, hard ride. He will carry you farther. Horses are ideal for taking you far around the enemy's rear and breaking up troop concentrations by swiftly moving surprise attack.

Make a reputation for appearing unexpectedly from nowhere, striking hard with lightning quickness, then vanishing as mysteriously as you came.

An added strength which comes with your mobility, is that your numbers are unknown. You are also stronger in that you are here today, gone tomorrow.

And remember that the enemy's Intelligence Service cannot keep tab of and act against you under such circumstances. To act against you with clock-like precision, Intelligence needs definite figures, positions, movements, and strengths. But you are indefinite, intangible. Hence they cannot plan to manoeuvre you into a position whereby they can bring definite and systematic force to annihilate you. Intelligence works in general to stabilize. But you are elusive. Furthermore, today you are only twenty men, tomorrow you may be 500. You are not confined to roads; you are *anywhere.*

You may appear to their front or flanks or rear; by night or day. You are phantom men. Your leader should regard this instability as an added weapon to defeat the enemy's Intelligence Service, and through that service the enemy.

Think upon innocence, then act upon it, for it can be a continuous ally, and a deadly menace working for you. A forlorn, abandoned hut can conceal a machine-gun. Who'd have thought it! A saltbush can conceal a sniper, a jam-tin a grenade. A rosebush is innocent enough, yet it can harbour a rifleman; the path beside it can disguise a mine. You will kill far more enemy by subterfuge than by tackling him like a bull at a gate.

If you find that the enemy are stalking you or stalking your hide-out, don't turn on him immediately, unless circumstances demand it. Take the opportunity to lure him on; then set some of your best stalkers to stalk him. Follow up his rear, so that when you do turn on him none of him will get away.

If the enemy develop the habit of chasing you in car patrols, choose their route one day. Lead them well away from help, out into the hills, the latter part of the journey into country which will force them to return the way they came. An ambush will be awaiting them at the most awkward place of course.

Whenever possible, make the enemy waste his petrol. If you can lure him out into the bush and make him do that, he becomes a sitting shot. Shoot him up at night if you are not strong enough to finish him off by day.

Don't forget that you can use dummy planes, dummy traps, dummy mines; dummies to distract his attention, dummies to draw fire. "Things are not what they seem", is a true saying. Turn this fact against the enemy. Remember that unexpected fire into the enemy's flank or rear is far more deadly and demoralizing than fire against his front. Always plan to do the unexpected and deadly things to him. If anything goes wrong, don't loose your head. Immediately strike the enemy from another angle and thus place *him* in the wrong. If you are operating in enemy country and see a nice stretch of road, don't walk up along it. You'll stop a shower of bullets if you do. And don't expect that the enemy will necessarily walk straight up along a road. If possible, he will very likely travel along each side of it. Only when he has occupied the road is he likely to openly travel along it.

All the above contingencies would be ruled by circumstances of course. But the enemy is cunning, and he also, despite the fanaticism of the minority, values his own life.

There is a very nice road not so very far from where you are now. At each end of it is a settlement of ours. The enemy lands, wave upon wave of men. Do they rush up along that road? Not on your life. In their rubber-soled shoes they swiftly marched up along both sides of it, well spread out. They thus outflanked the reception committee awaiting them. beyond range, or to the second cover which is within range.

You see, the enemy is very shrewd and will not poke his head into a trap unless it is excellently baited. So, think out the right bait. And do not fall into a mug's trap yourself. Never let a problem beat you; keep at it until you think a way around it. The solutions to many problems are laughingly simple - when you think the right way.

Now, we'll think out a problem. You are bike men, operating in bike country. Before you is an unsurmountable range, or a line of the enemy. It appears utterly impossible to get bikes over that obstacle in front. And yet, you know that on the other side of that obstacle, say thirty miles away, is is good bike country. Another twenty miles farther on is an enemy camp about which you have full information, and which you are confident you can wipe out.

Impossible. You cannot get bikes over the range, or through the enemy in front, let alone make the attack and the getaway right back to here. Well, send details of the whole business to your nearest Military Headquarters. Ask the Heads to loan you a transport plane or two to land you and your bikes over the obstacle, and to pick you up when you return from the raid.

That is a simple illustration of how to use your wits. Never let an opportunity slip.

Beware the hidden sniper!

Never fight a losing battle. Plan for success before attack. No guerrilla band should be a static force. It should be "elastic" in its composition, its life, its fighting, and all its movements. The only "solid" thing about it should be the "solid elasticity" of its attack. In other words, the principle of guerrilla warfare should be that the guerrillas strike solidly at the enemy's weakest point, but vanish like mist at his solid attack. Thus he can never corner them; never get his teeth into them.

Be resourceful, always brimful of initiative. Your best ally will always be your head. For your weapons of morale against the enemy use surprise, fierceness, distraction, confusion, decoy, diversion, ambush, feint, irritation. Wear them down. Pretend to strike here, but attack there.

Break through their weakest link and leave sleeplessness and the terror of the unexpected behind you. Never let the enemy know of your hide-out, Change position again and again. Be secretive in movement, unforeseen in action, vanish like shadows. Use mobility in concentration, attack, dispersal, uniting. Make your scout system and communication between you and other guerrilla bands and our own Military, far superior to the enemy Intelligence. Combine for mutual protection, for defence and attack.

If you are operating in a district which for years has been thrown open to enemy aliens, it is almost certain that numbers of enemy sympathizers there are not interned. Therefore, use your eyes. Does anything attract you about the houses, the farmhouses? Is the washing on the line present more often than it should be? Does one garment continually appear, and enemy planes appear shortly after? Does smoke rise from the chimney at odd times? Does it rise when there seems no need for it? Is the dog chained in one possy today and in another tomorrow when troops are passing that way? Are ten white fowls in the farmyard when ten of our own planes are stationary away out on the bush drome?

How is the farmer ploughing? As usual lengthwise, or are his furrows crosswise or diagonal, maybe converging in a certain direction? If so, are those converging furrows pointing to our artillery positions miles away? or to our drome? or to our hidden park of tanks, or horses. Does the farmer plough in a nice white shirt that can be seen many miles away from the air? In various districts in Australia these questions, and many like them, should occupy your eyes and mind. Remember that there are numerous ways in which an ingenious enemy in our midst can give signals to the enemy quite apart from wireless and other well known forms of signalling.

If in such a district, keep your eyes open lest an innocent farmhouse be signalling in most innocent ways "direction" to distant enemy artillery, let alone to enemy planes. Unless necessity demands it, never expose yourself until the coast is clear.

Surprise, speed, and attack win modern battles. Apply this modern principle to yourself. With surprise and speed attack boldly and you will win out. In the last two years empires have been won with speed, surprise, and attack. Use these mighty weapons yourself. Remember that, due to mechanization, many modern military movements are actually run to a time-table. Smash one cog of that time-table and--! Hence, it is vitally important that you smash that convo-y. If you can do it, you break the enemy's time-table. He, or his supplies, or something vitally affecting him, arrives late and our regular Military forces win the battle.

Observe, without being observed.

Always be the hunter; but be careful lest *you* be hunted. If this happens, circle and come up behind *him*. *Always* be the hunter.

Remember decoy fires. Small "carefully" concealed fires deep in the bush. The enemy sees them; creeps up to surprise your camp. You surprise him.

In dealing with parachutists, remember that very probably you outrange them for ten to fifteen minutes after they land. Their machine-guns are seldom within reach until then. On first landing, the majority of parachutists who have gone into action so far in this war have landed with automatic weapons. The ordinary military rifle far outranges such weapons.

If the enemy wants your position in a furious hurry, then let him have it - at a cost. There are millions of good positions in this country. Make the enemy pay a price for each one. In the end, he will own none.

Make every movement, whether for attack or defence, for a definite purpose.

Develop confidence in yourself. The guerrilla who it confident he can do the job almost certainly will do the job.

Put yourself in the enemy's place. Reason out what he would expect you to do, then act quite differently.

Don't forget that shade and shadow is really cover. Slip away from the shade, though, and your own shadow might betray you.

Concealment is vital. You have the enemy beaten if he cannot locate you, or your position.

Thoroughly learn every nook and cranny of your own district. Not only will this give you a priceless advantage over the enemy but will make you invaluable as guides to our regular army.

Learn the quantity of food necessary to keep you fighting fit for a hard week in the field; so that, when you are sent away on a job that will keep you a week from camp, you know what food you need.

Learn what time means. How long does it take you to ride twenty miles? Or to walk twenty miles? All things being equal, how long will it take an enemy truck to travel twenty miles along a given road? Only by learning the actual meaning of time and distance can you plan and successfully act to intercept an enemy, or plan attack and getaway where time and distance is concerned.

Remember that time and distance have a different meaning at night. The time is the same; the distance is the same; but it takes longer to travel it than by day. Particularly would this be so if you were guiding troops over a rough, unknown route by night. Those men would not only travel slower, they would find the "going" much more difficult than under the light of day. Remember little points like these, for you may be called upon at any time to guide troops through country strange to them.

Develop your eyes; learn everything you can; constantly

seek to better your judgment. For instance, you may walk or ride for miles along a bush bridle-path. Could *you* then guarantee that that path is negotiable for our military lorries or tanks? Such a question might arise on an emergency and if you had not used your eyes and judgment you would not be able to answer. The officer commanding may urgently need to know if there is cover for his troops along a certain route; cover for his guns; if there is water; the distance between water and water; if there is cover from the air as well as ground cover; if there are any points along the route where the enemy could lay an ambush - and so on. If you cannot answer these and similar questions, it will mean you have not used your eyes and judgment.

The good God gave you brains to use. Use them in the service of your country-women and children, your own mates, and your country. If every man did that, nothing on earth could stop us from easily winning this war.

March 1942.

11

Putting the Shoot into Parachute - Danger Zones of Fire

IN the last chapter we very briefly mentioned the arms of parachutists. As you will come up against them we'd better say a little more about this subject.

Parachutists are not dropped at random; they are dropped in a definite place, and have a definite and immediate duty before them. Firstly, enemy reconnaissance planes take photos of the country over which they propose to advance. One photo shows an area which would apparently make a good landing-ground; it is in a strategical position favourable to the enemy, and the photos show that it is not heavily defended.

At a chosen moment the first parachutists appear. Their immediate job is to seize that ground and make it safe for their reinforcements. Wireless then brings the troop planes and in a very short space of time that area has been seized and consolidated by the enemy.

There are quite a number of other objectives. They may descend to seize a power house, or railway station, or drome; to disorganize a vital line of communication, or open unexpected fire into the rear of one of our important positions. Always the parachutist has an immediate objective before him. He is a highly trained and skilled man and his energies are not thrown away.

It is very necessary, then, that paratroops should be destroyed before they have time to take and hold a position. Otherwise their troop planes may soon land thousands of the enemy.

Don't forget that to defeat parachutists you must use your brains. Although they come from the air they use the old ground tricks. For instance, if you were watching an area that our Military wished to be held against possible paratroops, and if the enemy wanted that area very badly, they would, as likely as not, sacrifice a few parachutists by obviously landing them a couple of miles away. If all or most of the band rushed to the spot, the main body of parachutists would descend on the ground you had just vacated. They are up to all sorts of tricks, so don't for a moment imagine they are easy prey

The best time to tackle parachutists is as they land. You must be on the spot of course. We'll see how the parachutist lands, and the best way to get at him. He is more or less, but not altogether, helpless as he comes floating down. The fleeting seconds keep him very busy, for he must handle his parachute; he is encumbered by clothes that are especially arranged to help break his fall; he is moving swiftly down to strange earth where he knows not what is awaiting him. But he is armed with automatic pistols, grenades: are in his pocket, he carries a long knife.

He probably lands with a jar, and quite possibly may sprain his ankle; he is flurried; he may or may not receive a nasty bump; he is quite helpless for a few moments after landing; he must disengage himself from his parachute, then fix his clothes. And so, he is in no position to fight while very hurriedly engaged in doing all this. Minutes are precious indeed to the parachutist - and to you.

Floating down with him and his mates are other parachutes. These carry long metal containers, in which are his machine-guns, mortars, anti-tank guns, grenades, explosives, special tools and ammunition.

The parachutists land all over the place, work frantically to disentangle themselves from their harness, then rush to where the containers have fallen. Swiftly they open these and hand out the weapons. Thus they concentrate and rapidly equip themselves. If smart work (and it almost always is), the time taken will have been only about fifteen minutes. Now they are organized, and armed with machine-guns, mortars, in addition to those already described-a far more formidable body of men. If you are not already tackling them, they carry on with their job.

All the same, despite their specialized training and very modern armament you still have quite a lot on them. You know the country; you are hidden; are. good shots, and your rifle has the same range as their machine-gun. So shoot straight, for these highly skilled men are also very desperate men. Mark out those containers-the enemy must rush to them to obtain their weapons. Try to beat them to it; then turn their weapons against them. Get all around them. If they have to fight their way through, you might wipe out the lot of them. If not, fight to hold them up. Thus they cannot attain their objective, while sooner or later help. must come to you. Whatever you do, don't lose sight of them. Because that, above everything else, is what they want you to do. If they can only fool you, draw a red herring across the path; if some of them can lead you on a wild goose chase, then the others can quietly slip away and carry on with their job.

You must be up to date. These facts give you an idea of what paratroops can do, or can do if they are not halted.

They can land tanks by parachute, up to ten tons so far.

Planes may already be specially built which can land heavier tanks. By parachute they can drop machine-gun carriers. They can transport trucks, anti-aircraft guns, very considerable heavy equipment, and infantry reinforcements to back up the paratroops. Those transport planes are especially built of course. But the knowledge of what they can do gives you an idea of what a danger unchecked paratroops can be.

Such troops, for various reasons, would be a considerably greater menace to a heavily populated country, than to Australia. Although they would be a serious menace to our cities and armies.

Now you'd better know the points that count with the latest automatic weapons: sub-machine guns, tommy-guns, etc. Their accurate range is very short, but very deadly. Instead of one bullet they send a stream, and that stream need not even be aimed-not as we aim with a rifle. It can be fired from the hip, almost from any position. This means that, besides being deadly they are very quick. At close quarters your best weapon against them is to be unobserved. That may sound silly. It is not. If the man with the automatic rifle cannot see you, his weapon is useless. If he catches but a fleeting glimpse of you, or if a sound gives him an inkling of where you are, he sprays you and your position with bullets. So long as you are unobserved, you have the chance of the draw on him. At medium or long ranges, of course, your rifle has him at your mercy.

At close range a running man is a sitting shot to an automatic rifle. At close range, too, one man with a modern tommy-gun can, while his ammunition lasts, keep up a up a stream of fire which equals the fire from forty men using ordinary rifles.

There are various kinds of automatic weapons, some of which are more powerful than the tommy-gun. Some fire from a container, magazine, or belt; some hold more rounds than others.

Your guerrilla band by now should be a specialized, well-organized body of fighting men. Make no mistake about it, well-organized guerrillas quickly make formidable fighting men.

There are a few more advanced points for you to grasp. But always remember that, no matter what you know, you will be learning almost daily when the fighting comes. Only when you perfectly understand your weapons, the work they can do and the work they cannot do, will you become thoroughly efficient not only individually, but in teamwork.

You can, now, use your rifle very effectively as an individual. But there are a few points about the "effects of fire" which you should know in team work. When you are firing at the enemy, and the enemy are firing at you, neither is hit unless in the zones of fire - except by "accident".

To hit the enemy you must be a marksman and have the correct range. And vice versa. And you can also be hit though an enemy has not aimed at you, if you are in a danger zone. Now, where the bullets strike the ground is called the "beaten zone". Imagine that a hundred of you are firing at the enemy who is at a thousand yards range. Where your bullets are falling is known as the beaten zone. If some of the enemy were standing up in front (they would not be such fools) they would be in considerably greater danger than men lying down behind them. Not only because they would make such easy targets but because they would make the "danger zone" so much longer. A danger zone would thus be created at the height of a man's head to the heels of the farthest back man who was lying down in the beaten zone.

Do you get the idea? The bullets are aimed at the enemy.

They begin to fall when near the target. Immediately a bullet falls to the height of a standing man's head that man and any others on a line with him, right away back to the feet of the farthest lying down man, is in the danger zone. If the standing man is not hit, the bullet carries on, rapidly falling. There may be a man lying down a fair number of yards behind the standing man, but the bullet will come to earth in the man's heel. That is the limit of the danger zone. Men standing out in front of the first man, or lying down behind the rearmost man who was hit, could not be hit by such a bullet or bullets. All the men, however, who were in between the two, whether standing, kneeling, or lying down, would be in danger of being hit.

Now the centre of that line (or rather patch of ground) where the bullets are falling would receive about seventy-five per cent of the bullets. And this particular area would be the most dangerous to the enemy - or to you if they were firing at you. This particular portion of the beaten zone is called the effective beaten zone.

This teaches you that if you conceal yourself well, you can probably defy the enemy getting your range. Hence he can fire away as long as he likes but cannot get you if within a danger zone, let alone an effective beaten zone. This helps to explain why it takes a ton of lead to kill a man. If he is not in a danger zone he is only killed by an "accidental" shot, a "stray" bullet.

A danger zone can vary and considerably lengthen, according to the nature of the ground. That may slope or fall more or less steeply, and it may slope nearly parallel with the falling bullet; if so, the danger zone considerably lengthens. Put it this way: Say you are firing at men concealed on the steep face of a hill a thousand yards away. There will not be a broad belt of danger zone where the bullets fall, because they strike straight into the face of the hill. You must shoot straight to get the enemy, for in such a position their men are in a very "narrow" danger zone.

Now imagine firing at that same hill, with the enemy replying from the crest of the hill. Behind them, the hill falls away on a sharp slope. You still would have to aim straight to hit any of the enemy, because the bullets would still plunk into the steep hill crest. But numbers of bullets would fly a little bit high and whizz just over the crest. Now, these bullets would create a broad and dangerous zone of fire - not to the men they were aimed at, but to any of the enemy lying down just over the crest of the hill, and also to any supports who were coming up the hill from away back.

The reason for this is the fall of the bullet and the steep fall of the back of the hill. The bullet is aimed at a man on the crest. That bullet is already steeply falling. It grazes the crest and whistles down the other side. According to the fall of both bullet and hill that bullet may travel the height of a man, or less, for some hundreds of yards before it actually strikes the ground. Hence, any man in its line of fall from just over the crest to where it strikes the ground is in extreme danger.

That little illustration in zone of fire should teach you something. If ever you are approaching such a hill (coming up behind it), while some of your mates are under fire on its crest, you will know there is a danger zone coming down the back of that hill. If you walk into it you very probably will stop a bullet. Hence, approach from right or left of the line of fire.

Also if *you* are doing the firing, and your scouts report enemy reserves on the back of the hill, you can very probably put them in a danger zone, even though you cannot see them.

Learn all you can about bullets, for there is a great deal more to them than merely shooting them from a rifle. If you understand the laws that govern bullets in flight you can in many an engagement tell at a glance the danger zones to friend and foe, and the danger zones which will occur if the enemy comes into action from such and such a direction and firing at you in a certain position.

This knowledge may come in mighty handy to you at any time. You can tell for instance whether there is danger over the crest of a hill facing you or behind you. A day will come when you intend to attack a position. You vitally need safe cover for your horses or your reserves. If you understand the laws governing bullets in flight, you will not shelter either horses or men in a possy where (allowing for the unexpected or swift changes of position, etc.) they would come under a zone of fire.

To illustrate again, very simply: You are a band of mounted men about to attack a detachment of the enemy in plain view on a ridge a thousand yards distant. Your ridge gives you excellent cover. You have dismounted and are selecting your firing positions. Your horses are behind you, at the foot of the ridge, and invisible of course to the enemy.

Now, those horses may be quite safe, it depends on the shape and fall of the ridge. On the other hand, every enemy bullet, soon to be aimed at you, may be a danger to your horses; for the fall of the ridge may be such that every bullet coming over it may fall right in amongst your horses. Thus your mobility will be cut to pieces, and you may fall an easy prey to a mobile enemy force.

12
The Danger Zone - Don't Be There

DANGER zones vary in length and depth, according to range; the closer the rifle to the ground-level, the higher the target, and the confirmation (the shape or slope) of the ground on which the bullets fall. If the ground were level, the farther the range the less would be the depth of the beaten zone; that is, the patch of ground on which the bullets are falling; because, the greater the range, the higher the bullet must rise, and the more steeply it falls. If the bullet travelled straight, flat-out as it were, the result would be different.

At two thousand yards, bullets fall very steeply indeed; a little less steeply at a thousand yards. We'll say that the ground is flat where the bullets are falling. The enemy are lying a thousand yards out there, firing at you. Each man will thus be a foot "high". Well, because of the trajectory (the curving fall) of the bullet, the dangerous space is only a length of ten yards. (The trajectory of the bullet will be one in thirty.) Therefore, all the enemy who are one yard nearer you than the edge of that dangerous zone, or one yard behind the zone, will not be hit.

Now, you should begin to understand what may really lie behind the saying: "It was a miracle how he survived. The ground was literally swept by fire and he was not hit!" It was because "he" was actually out of the zone of fire. And that other remark: "You could not imagine how any living thing could possibly survive in such a hail of bullets!" Again, it probably was because the "living thing" was just out of the actual danger zone.

These "miracles" happen continuously throughout every war. In four years of it I saw great numbers of men go through zones of fire unscathed. But they were the ones for whom no bullet carried a number. You may be pouring in a hail of fire and yet not hurt the enemy unless they are actually within that shallow danger zone.

Realize, then, the extreme importance of quickly picking up the correct range.

We will now see what happens when a body of men fire at the enemy six hundred yards away. At this range the bullets do not rise so high nor fall so steeply. Hence, the danger zone is deeper - thirty yards. (The trajectory of the bullet will now be one in ninety.)

Hence, you might be able to put it this way - the ground being equal: that it is three times more dangerous to be fired at from six hundred yards than it is from a thousand. And less again at fifteen hundred. Say that you are in supports, and ordered at the double to the firing. line. It is already under fire, the range a thousand yards. You will be comparatively safe until you arrive at the edge of the danger zone, except from "highs" and "strays"; in fact, until right up among the falling bullets.

We are imagining that the line is open ground, under correct range fire. You crawl into the danger zone. The next ten yards you are under fire. Every inch you raise your body you greatly increase your danger. You are in extreme danger all the time; not necessarily from an individual enemy aiming at you, but from the hail of bullets falling into that danger zone. Thus you crawl forward another ten yards; then a few more yards - and you are out of danger, except from strays, or bullets falling short.

To do that at 600 yards (conditions being equal) you will have to crawl thirty yards through the danger zone. A few yards further and you will be comparatively safe, unless you stand up. If you do that you instantly bring yourself up into a danger zone.

The above illustrations should give you an idea of zones of fire; teach you that bullets are governed by laws; that there is a limit to their dangers; but that they can be made far more dangerous by the man who understands and directs those laws, than the leader who does not.

The closer you are to an enemy the more dangerous his bullets, not only because he may see you and thus aim, but because the bullets are coming "more at the level". In other words, the trajectory is flatter, the bullets travel much closer to the ground. When you sight the rifle your eye looks at the target in a straight line, but the bullet rises towards it on a curve, to rapidly drop towards the target. When a number of men are firing the flight of the bullets is known as the cone of fire. The closer the range, the closer is this cone to the earth. Thus the shorter the range the more rapidly the danger zone increases.

That cone of bullets should give you food for thought. You may be only a few hundred yards away from the enemy and invisible to him. He cannot see to aim at you. Still, your mates are getting hit. This is because the danger zone is deeper, hence more dangerous with the shorter range. You are really now under this cone of bullets; it is not so much that individual men are firing towards you as the fact that you are lying in a dangerous space produced by the fall of that cone of bullets. Something, or someone amongst you must have put you away to the enemy. He does not see you but he has your range to a nicety and has brought you under a danger zone.

That is one more reason why you should take the greatest precautions against any indiscretion which may betray your position to an enemy.

We have just explained that at long range, if the ground falls sharply behind the enemy, then the danger zone is considerably increased. At shorter range a similar law applies if the ground falls *gradually*. This is because the bullet at shorter range is falling much more gradually than it would be at longer range, and any of the enemy coming up behind their firing line would be liable to stop a bullet.

The ground, of course, must slope parallel with the fall of the bullet; for immediately the ground rises the bullet strikes the earth. If the ground falls steeply the bullet whistles on. Hence the shape of the ground, in relation to fire effect, is very important indeed.

Think of that when the time comes for you to choose a position from which to ambush, attack, defend, or put up a delaying, or running fight. You may consider a hill or a rise as a possible position. Study it first. Does it slope smoothly yet steeply to what will be your rear? If so, it means that when the enemy comes, then from long range he will be able to put you under a lengthy and particularly dangerous zone of fire if you eventually retire down that hill. It means also that reinforcements coming up to you must come through that long danger zone. If you are horsemen it means that your horses will not be safe away behind the hill.

You look about for a better position.

If you understand the danger zones of bullets you can look at a position, then glance around at positions which may be occupied by an advancing enemy, and in your mind's eye "open" fire on the position you are thinking of occupying. In that way you can get quite a good idea of the areas the enemy could or could not put under a dangerous zone; and you could estimate the length of the zone. So, you will not unwittingly take up a position which can be swept by a long, dangerous zone of fire.

If possible, avoid a position on which there are large patches of bare rock. These are very dangerous; for bullets ricochet at all angles and would be liable to glance off that rock and shoot away for perhaps hundreds of yards. Quite often in such country, a bullet may strike a rock and ricochet for a hundred yards only to strike another rock and ricochet again. It may do so three times; even more. Thus its length of "danger life" is considerably prolonged. Not only so, but ricocheting bullets are nervy missiles. Then again, bullets striking a bare rock surface, splinter and splash, and many wounds can be caused by those splinters. Further, they break off splinters of rock which can easily blind a man.

Remember, also, that if you were brought under artillery fire upon a bare rock surface your casualties would become much greater. The shells would smash into a thousand splinters of flying steel, and carry a thousand splinters of rock with them.

Your guerrilla band is composed of a number of rifles. To a considerable extent, your strength lies in the fire power of those rifles. That fire power is your "punch". If that punch is a hundred per cent effective, you are very strong. But your fire power may be only sixty- even forty-per cent. Still, if your marksmanship and judgment of distances is good you can make your fire power one hundred percent, provided you understand the laws that govern the flight of bullets. Not unless.

The elevation of a bullet, the rise and fall from the moment it leaves the muzzle until it strikes the object, explains why it is possible for men to advance a thousand yards under heavy fire and still live to tell the tale. If well led, they are only a short, time in the actual danger zone - until, in fact, they are at fairly close range. Most of the time the bullets are flying over them. The approaching soldiers say to themselves: "Thank heaven! They are flying high. May they continue to do so."

Really, they are flying high because the men firing them, growing more and more excited as their enemy comes into plain view, forget to keep lowering their sights. Hence many of those men advancing to seek the riflemen's blood will presently be right up into them with the bayonet. They could never do it if the officers commanding the defence understood and controlled their fire. By doing that the attackers would be compelled to advance every yard of the way through one continuous danger zone. Not one man would get through.

It is very important that you should understand this, both in offence and defence. Practically applied, the knowledge will make you many times more formidable. You will coolly and continuously direct and control your fire power, and be able to take advantage of an enemy who does not understand how to do that, or who has become too flustered and excited to remember.

When a position must be taken at all costs the final job generally falls to foot troops. Tanks may smash into a position, but infantry is needed to clinch the capture, to consolidate and hold. If the enemy land in great force in various parts of Australia there will be many a fight of infantry against infantry. If mounted troops cannot take a position at the gallop, they act as infantry. Whether you are mobile men or footmen, you will again and again be infantry. And probably will quite often be called upon to fight without the help of tanks.

Artillery and planes bombard the enemy position, thus giving you covering fire.

Probably you will be ordered to advance, in scattered groups, to get into the flanks of and around the enemy, to strongly infiltrate into them from all sides if possible, while taking every advantage of cover.

Now, we'll attack an enemy position by the old method of "short rushes", and we'll be out in the open. You may get an inkling then of some of the meanings of fire power.

There may be a line of you a mile long. The whole line does not advance at once, but by sections. Thus, a company may leap up, run ahead a hundred yards, then drop to the ground and open fire. The companies nearest on right and left remain flat, pouring a brisk fire towards the enemy's position, in order to keep their heads down. The enemy's aim has been disturbed as they fired at your advancing mates. The two covering companies now leap up, advance, and drop, to again fire. Thus half the line with gaps between leap up; advance a hundred yards at the short run; flop down; immediately get the range, and open up a furious fire towards the enemy. Then, and then only, the companies that had covered them leap up and run forward in their turn, to flop down and reopen a covering fire. While they gain their breath and this fire grows quickly in volume, the first companies who had advanced, leap up and run forward again. So it goes on until the complete line is within two hundred yards, or less, of the enemy. Then the order is given; bayonets flash, and the complete line rises up and charges.

Now, here is where you must understand the zones of fire. At the beginning of the attack the defenders got the range. The attackers were caught within a danger zone of bullets, and men began to fall pretty fast. But, when those first companies leaped up and ran forward, the range so far as they were concerned was immediately altered, a difference of one hundred, or two hundred yards, or two hundred and fifty, as the case may be. Hence, instead of running into fiercer danger, they had actually run (when they flopped down) pretty well out of it.

Thus danger zone, so far as the advanced men were concerned, was altered, not by the enemy, but by the advancing men themselves.

To counter this, the defenders had to act swiftly, coolly, and "scientifically". They had to bring those advanced men into another dangerous zone of fire. Let us see the difficulties the defenders were up against; and let us see whether the attackers use their wits.

FREDERICK, MD., MONDAY, JUNE 1, 1942.

Japs Try To Raid Sydney; Three Midget Subs Sunk

Thunder Of Gunfire, Depth Charges Greet Foray By Enemy In Southern Australia

Sydney, Australia, June 1 (P)—Japanese midget submarines, apparently launched from a mother ship off the Australian coast, sneaked into famous Sydney harbor

Bulk Trapp

Escapee O₁ Trial For R At Hagerst

Anderson Sane,

Tw Dwantan Sau

131

13

The Danger Zone - Keep Them There

IN the first place the defenders have seen this start of an attack which now plainly means one thing - a fight to the finish. That very knowledge immediately thrills every man of the defenders; no matter who he is or what position he holds, he grows in greater or lesser degree a little more excited. And excitement dulls the memory. Besides, shells and bombs are falling around him, explosions and flying earth; and a stream of bullets, machine-gun and rifle, are pinning his head close to earth. Often dust, smoke, flying gravel prevent him from shooting straight. Moreover, there are large gaps in that advancing line-now really two gappy lines, one ahead, the other behind. Hence, two ranges. "Which will he fire at? And what is the range now?" Some among his comrades do not even think these questions; already they are beginning to fire wildly.

No sooner do the officers shout orders to counteract the change in range of one-half the advancing line than the other half springs up and races forward. A difficult target because the men are running, and because of those confusing gaps. Many of the defenders excitedly swing their rifles to right or left to get a bead on one of those advancing companies and by doing so lose all "touch" with the line they were previously firing at. And while they hesitate their range is again altering with a new advance. Many of the defenders forget this and don't lower their sights, with the result that soon this second line of attackers has advanced out of the danger zone. Again, they have created yet a different zone to the men who first advanced.

Just as the defenders get the range of the first line and put it under a danger zone the men rise up and run forward; again more or less dodging the danger zone. And so the game goes on to its final, breath-taking conclusion.

The attackers' object is to run forward to cover ahead, if any; dodge the danger zones; keep both lines of men "intact" and under control, and with their wind as unimpaired as possible, so that all will be comparatively fresh for the bayonet charge.

I hope you understand now the "danger zones of fire", and also understand how necessary it is for you to keep an enemy under a continuous danger zone. Once let him get out of it and you are only wasting your time and bullets until you get him into a new danger zone.

I have seen men advance on, more than one occasion across 2000 yards of level ground that had no cover of any description, not even a blade of grass. It appeared sheer suicide to attack heavily defended, strongly prepared defences in the form of redoubts set upon low hills in the centre of a plain. Those big rings of "wolf" pits, gun-pits, trenches and lines of redoubts commanded all the plain completely around them. The attackers were in plain view. They attacked under aeroplane, artillery, and concentrated rifle and machine-gun fire.

It seemed certain suicide. Yet the positions were taken, and at miraculously small loss. The secret was, of course that those attacks were so well handled that the men were continuously dodging the dangerous zones of fire. They were even helped by the enemy who, becoming more and more excited as the advancing regiments came nearer and nearer, often completely forgot to lower their rifle and machine-gun sights, let alone continuously seek the constantly changing ranges. If ever you are defenders and act similarly then the enemy will get right at you with the bayonet while you are marvelling why you cannot kill them.

Under most Australian conditions you would not be called upon to advance over such a perfectly flat, utterly bare surface. There would almost certainly be cover of some sort. Also, it is very improbable that you would advance in definite waves - in short rushes. You would scatter into many small, irregular groups or units and advance individually in short rushes or bounds, springing up here, there, everywhere; each group acting as an "individualistic group", while yet acting in team work. That would still further throw the enemy's fire power out of control. Instead of having definite waves from definite lines to contend with, he would be harassed by hundreds of small groups apparently having no cohesion at all. You can imagine the confusion in his ranks while endeavouring to get the range of all those scattered groups, and keep them under his fire power.

It could not be done.

A standing man, of course, creates a danger zone for himself. The range may be such that he can walk, while the bullet flies over his head. On the other hand he will eventually walk into a danger zone much sooner than the man who is kneeling, much sooner again than the man who is lying down. Look at it this way: Say a six-foot man is standing up. We'll imagine his head is one foot long. His length in "heads" would be six. A man is lying down beside him and his "head" is only one "foot". A bullet comes, dropping rapidly. It is not aimed at either 'man, but the "top head" of the six-foot man is just in its falling line of flight. It "knocks" that head, whereas the head of the lying man, five heads below, is perfectly safe.

When you are advancing against rifle and machine-gun fire you will double yourself up into the smallest possible space. You will imagine you are making yourself a smaller target for the enemy firing at you. So you will be; but you will be doing something else; you will be "lowering" yourself from the danger zone of fire; not from the man you imagine is firing at you, but from those many hundreds of vicious bullets whistling all around and but inches above you. When you lie down, you lessen the danger zone considerably.

There is an illustration which should show you a cone of fire, a nucleus, an effective beaten zone, and a dangerous zone. Imagine a man with a garden hose and a large garden; from where he stands he is watering the extreme end of it. The nozzle of the hose is in his hand. But the nozzle points up in the air-points exactly as a rifle-muzzle would point. The stream of water shoots from the nozzle and curves up into the air to come down in a steeper curve. Well, that curving water streaming up and out from the hose is your cone of fire. It is all the bullets being fired by say a thousand men at a target a thousand yards away. Just before the steeply falling water hits the ground you notice the volume spreads out a bit. Well, bullets fired from many rifles spread out very similarly.

Where the water finally hits the ground is one big blob of water. That is your nucleus, where most of the bullets are falling. Death is there.

As well as making the blob, the water splashes in a more or less lengthwise puddle. That is the "effective beaten zone". Very dangerous to be lying there.

Beyond this puddle are little puddles and plenty of splashes. They form the dangerous zone. A man lying anywhere within that area of falling bullets might be hit. If he stood up, he probably would be hit.

Outside the "dangerous zone" of the splashing drops of water you can see that a man would be safe except from stray bullets (note the occasional splashed drops beyond the zone) and bullets fired from some wildly shooting rifleman. You can understand, too, that a man midway between where the. water falls and the man with the hose, that is, in the middle directly under that stream of water. (the cone of fire), would be quite safe.

Now, the man with the hose wishes to water a flower-bed which is nearer to him. He turns the nozzle in that direction but he lowers the nozzle. So you must lower your rifle sights. The water from the hose now falls on the nearer flower-bed, and that the water from the hose (the "cone of fire") curves less in the air than it did at longer range. Hence, a man standing in the centre of its path would not be quite so safe now. You notice, too, that though the water in falling makes quite a splash, it does not water any plants which lie outside the radius (danger zone) of that splash. To water those plants the man with the hose has to alter the direction and elevation of the nozzle.

Regard the plants as troops, and you can see that they would not be hit unless brought into a danger zone of fire. You can plainly see, also, that if you can confuse the enemy, or outmanoeuvre his dangerous zone of bullets, you can work "under fire" and still not be hit.

14

Confidence Wins Battles

ONE more weapon you should understand - machine-guns. For you will be up against them.

Given favourable circumstances, and operated by experts, the machine-gun is the most terrible weapon of modern war. It is good at 2000 yards, deadly at a mile, while at close quarters it can be a slaughtering machine. As you will capture enemy machine-guns, learn what can be done with them.

They are nerveless. Infantry are men, and their shooting becomes erratic when they are fatigued, or excited, or hot and bothered, or cold or wet or miserable. Even the well-trained man does not shoot so straight and coolly if under a hot fire.

But none of those human weaknesses affect the machine-gun. It can swing to any flank, keeping up a continuous stream of fire. Whereas, if a detachment of you were suddenly attacked from the flank your leader would have to move men to protect it. The machine-guns supporting fire is of great intensity and volume; it can direct bracketing fire upon roads, communications, assembly points for enemy troops; can shoot a stream of overhead fire; bring to bear upon the enemy concentrated and collective fire or dispersed fire; swiftly spray danger zones for various ranges; and a battery of machine-guns can direct interlacing cones of fire upon an enemy. It can be fitted by day into fixed positions to open up deadly fire at night.

The machine-gun does not fear artillery if it can get within 2000 yards range, for then its spray of bullets can wipe out the gunners. It can be used in any type of country; can operate under practically any conditions that the troops can. It is not a large target, while its crew of from four to six men can pour in a volume of fire at least equal to fifty riflemen. In this respect alone, possession of only one machine-gun would make you much stronger. Gain possession of a dozen.

In a fixed position two men can operate a machine-gun. If one falls, a rifleman can take his place and the gun will still be worth fifty men. It means to the band economy of men, and much greater strength, as well as much greater danger to the enemy.

You can carry the gun easily on a packhorse and the crew can quickly run it into cover. The cover necessary to shield a gun and complete crew is not nearly as much as for one platoon of infantry.

Then again, the effect upon enemy morale of machine-guns suddenly opening up on them is very great indeed, apart from the damage they do.

As in everything else, machine-guns and their uses must be fully understood before the best results can be gained from them. The leader should, by a glance at the country, know where to place his guns and work them to the greatest advantage; should know tactics that will be employed against him; must understand how the enemy's weapons will affect him; in short, he should be an expert at his game.

A badly worked machine-gun suffers from the same faults as a badly handled rifle, and you know all about that. If the wrong elevation is on the sights, if observation of the enemy is poor, and the effect of the fire is badly observed and corrected, then the gun is just about useless.

You may not have received any machine-gun training; but by now you should know all about the rifle, cones of fire, danger zones and fire direction. Well, apply the same principles to the first machine-gun you capture and you will make a good fist of the job. Machine-guns are simple to operate, and your knowledge of the rifle and fire power will stand to you.

Now that you know how dangerous a machine-gun can be in expert hands, you will, take special precautions in tackling an enemy armed with them. The best precautions of all are - cover - surprise - straight shooting. While he cannot see you he cannot hit you. If you surprise him when his gun is in the open, you shoot the crew-quick.

As for the rest, if he is in position, infiltrate him, stalk his gunners. Don't give him a target. Advance under cover in small groups that split up into individuals. Pump lead into his position from all around him. Tackle him on the principle that he cannot fire in all directions at once.

And now, a final word or two about fire direction; it is so very important. To make the greatest possible use of your weapons and thus automatically increase your strength, you must plan, upon every occasion, to bring as many as possible of the enemy under your field of fire. You know all about zones of fire and danger zones. Well, here is another point:

Cross fire. In cross fire you get the enemy into a scissors; and the effect is particularly disastrous - for the enemy. Imagine a dozen of you are defending a stretch of road. A score of the enemy are coming' down the road; you open fire straight toward them.

If you are good shots and have the exact range, your twelve rifles straightaway bring the enemy into a danger zone. Now, each of your bullets either hits a man, or narrowly misses him, then ricochets away or buries itself in the ground. It has had just the one chance of hitting a man; and your target is only the blob of his head and perhaps a vague outline of shoulder.

Now, instead of you directly barring his way and firing straight toward him, plan to bring him under a cross fire. You expect him, sooner or later, to come along that road, and you get the exact range to a point where he will be most exposed; it may be 400 or 600 yards away. When he reaches that point, you intend to open fire.

Instead of lining the road, you choose two positions, one to the right, the other to the left of the road. Each position, of course, fully commands that point of the road on which you intend to open up fire. Six of your men, a nice distance apart, snugly fix themselves into the position on the left, six into the position on the right. Both parties are now looking from an angle at that point on the road. When the enemy reach that point both open fire.

What happens? The bullets from both parties strike that same point on a criss-cross; right where the enemy are lying bullets are both "coming and going", as it were. If a bullet misses one man it has a good chance of striking another near him. Not only so, but bullets are coming at him from his left front, and also from his right front. Further, practically the full length of each man is a target, and so he is in greater danger. Even ricochets are considerably more dangerous, for they' now must fly across the men.

Where the enemy is lying is the "rivet" that joins the open scissor blades. Your bullets are criss-crossing right on that rivet. The extended points of the scissors are your two positions.

Think about that situation; discuss it with your mates. You will realize how dangerous cross fire can be; and will plan to catch the enemy under a cross fire on every possible occasion - and take good care that you are not similarly caught.

Cross fire need not be confined to roads. It can be used anywhere: in the open country, on the sea, any position into which you are shrewd enough to manoeuvre the enemy. And you can apply it against any numbers. Also, the more there are of you the more "variations" you can bring into your fire: frontal, cross, flank, rear. Whenever possible, plan to fire at the enemy from all points of the compass, from every angle. When you do so he will fall quick and fast.

And now, as to dead ground. That is ground upon which, from a certain position, the enemy cannot hit you. If the enemy gets into dead ground, you cannot hit him. The laws which govern this really come under the zones of fire. You remember the angles at which bullets fall, the longer the range the steeper the slope of fall. Well, imagine you are firing across a plain; the enemy are advancing from a thousand yards; you are causing them numerous casualties. They reach 400 yards range, and their casualties practically cease, even though you have the correct range.

It is because they are now sheltering in and firing from dead ground; a little hollow or depression in the ground. Your bullets are not firing into that hollow because the angle of their flight takes them over it. So long as the enemy stay there they cannot be hit, unless they poke their heads above the hollow to fire at you. And then you will only hit them by a dead shot getting them straight through the head.

That is a simple illustration of "dead ground". When you lay an ambush, be sure there is no dead ground in front of you. Because if there is the surprised enemy will quickly occupy it and you'll have the devil's own job to blow them out. The only way you can overcome such a menace is to place men to flank or rear of it so that their fire, coming from another angle, shoots directly into that dead ground. To men so placed that ground is not dead ground. And don't forget the chance of a second trap here. The enemy, after their first surprise will rush that dead ground. Your men who have been especially placed to command it should not open fire on them until the, enemy are crowded into it. Then your men who command that ground, open up as strongly as they can.

If you have no means of commanding that ground from another angle, it will become a concentrating point for the enemy and a menace to you. So, when laying an ambush, very carefully study just where you intend to open up on the enemy.

Knowing there is such a thing as dead ground, you can be on the look out for it and take advantage of it, when attacking the enemy. Search your front for dead ground. Get into it. From there you may safely spy towards the enemy and see what your next move will be.

In addition to dead ground which may in part spoil your otherwise good field of fire, there may be a mound out there in front. Actually this is a form of dead ground which can partly shield any enemy who run up behind it. You can blaze at their heads if they show above the mound but you cannot put them under a danger zone unless that mound is at very long range. A house in your field of fire can also partly spoil the fire effect. Any obstruction, either a space below ground level or something above which prevents your bullets from hitting just there is a danger zone. Remember that a danger zone has length and width. Any man Iring in a danger zone is liable to be hit anywhere, from any bullet falling there whether fired at him or not. But if a man's head is the only target, and the range and conformation of the ground is such that when a bullet misses his head it flies straight on over him, he is only in danger from a crack rifle-shot.

For those of you who at any time may be operating on or near the coast or rivers, remember that boats can be a very serviceable means of dodging past the rear of the enemy at night, or of pulling around him to land quietly in his rear, or flank. Watermen of course will know all about this, so get in touch with coastal guerrillas if you wish to get right round the enemy and land twenty or thirty miles or so away towards his other flank, or to land behind him so as to infiltrate up through him. Many landsmen would be surprised at what small boats can do, and how far they can travel at night. Hundreds of miles, so far as that goes, by pulling into shore at dawn and hiding by day. Boats do not leave tracks. They do not make your feet tired through walking. They can carry a surprising amount of tucker, water, ammunition, and gear. With a very little ingenuity they can be fitted up comparatively comfortably against rain. Under various circumstances they make an easy getaway if your little raid is not far from where you've hidden the boat. They are ideal for landing venturesome scouts behind the enemy, in positions that could not be reached by land and from where he would never dream he was under observation.

Boats can be used in more ways than the landsman guerrilla may think. For instance, after a more or less long trip up the coast to get right behind, the enemy, a boat could be taken up a quiet creek; reserve foods and ammunition hidden ashore; and the boat quietly sunk just where you can find her later on. You are at liberty then (unless you fall foul of the enemy) to travel fifty miles inland if necessary, feeling quite fresh, well loaded with provisions, knowing there is a further goodly supply back near the boat. My mates and I have travelled in a whaleboat for months at a time like that, prospecting wild parts of the Australian coast. The transport problem for man, tucker, and materials is thus solved. When you land you are perfectly fresh for whatever your enterprise inland may be. If there are islands along the coast, a boat is still more useful. I see no reason why small boats should not be used by guerrilla fighters under suitable conditions.

You should know quite a lot about guerrilla fighting by this time. If you have understood, discussed, and memorized the hints and advice given in these little books, I'll guarantee you'll be unbeatable when the whips begin to crack. So, have every confidence in yourselves; it is the man who thinks he can do a thing who really does it. Napoleon and Hitler were only little corporals but they were simply bursting with confidence. They allied brazen cheek and grey matter to confidence, and you know the result. You have grey matter, too; use it and attack with confidence and you'll win everytime. Your grey matter will warn you when not to take an unnecessary risk; at the same time it will warn you when to dare. When that time comes, take the risk. Get in the king hit first. If after that things should go wrong, think clearly. Don't drop your bundle. Adapt quick action to the changed circumstances. Confidence grows when you realize that, even if you do make a mistake, it need not be your last. Think quicker than the enemy, and you turn defeat into victory.

Again - remember that unity is strength. Get in touch with other guerrilla groups, and with the nearest Military. Thus, you can not only combine when a larger operation makes concerted action necessary, you can swap information, and help one another in numerous ways for the common good

You could collaborate to simultaneously attack a strong enemy on several fronts; flank him, too, and cut his communications. You would thus quickly cut him to pieces instead of, possibly, being destroyed, band by band, yourselves.

Groups of bands could form a control council. This council would know the activities, strength, district, of each guerrilla band or group. All information would go through the council's hands. Hence, as opportunity arose, they could call or direct all bands in concerted action together. This council would also be in full touch with the nearest Military authorities, so that, for a big movement, you'd have their co-operation. After any such concerted action the band could separate to carry on as before, while ready immediately to re-unite when the word was sent around.

So long .. And good luck.

Japanese midget submarine being raised from Taylor's Bay.

ION 'Jack' IDRIESS was born in 1891 and served in the 5th Light Horse in the First World War. He returned to Australia to write *The Desert Column*, which was published following his huge success with *Prospecting for Gold*. He went on to write 56 books and was largely responsible for popularising Australian writing at a time when local publishing was still not considered viable. A small wiry mild-mannered man, Idriess was a wanderer and adventurer, with a vast pride in Australia, past, present and future.

ETT IMPRINT has published new editions of these books:

Prospecting for Gold (1931)
Lasseter's Last Ride (1931)
The Desert Column (1932)
Flynn of the Inland (1932)
Gold Dust and Ashes (1933)
Drums of Mer (1933)
The Yellow Joss (1934)
Forty Fathoms Deep (1937)
Madman's Island (1938)
Headhunters of the Coral Sea (1940)
Lightning Ridge (1940)
Nemarluk (1941)
Sniping (1942)
Shoot to Kill (1942)
Guerrilla Tactics (1942)
Lurking Death (1942)
Horrie the Wog Dog (1945)
The Wild White Man of Badu (1950)
The Red Chief (1953)
The Silver City (1956)
"Gouger" of the Bulletin (2013)
Ion Idriess: The Last Interview (2020)

ION IDRIESS
The Last Interview
TIM BOWDEN

Ion "Jack" Idriess (1889 – 1979) is recognised as one of Australia's great storytellers, having published over 50 books including the Outback tales of *Lasseter's Last Ride, Flynn of the Inland*, and *The Cattle King* alongside major works on the histories of Broken Hill, Broome and Cooktown.

This book is his last interview in 1975, prompted by the then young Tim Bowden, for a possible ABC Radio program that did not eventuate. With renewed interest in Idriess and his life, within this book Idriess talks of his early years in Broken Hill, he tells of his earliest writing for the *Bulletin*, on living and photographing Aboriginal tribes in the Kimberleys and Cape York; on the writing of his books like *Madman's Island* and *My Mate Dick*; his life with the pearlers of Broome and Thursday Island; on the joys of prospecting, living in the Wild, on Lasseter and his diary. Full of colourful characters and true stories, Ion Idriess allows us into his unbridled enthusiasm for Australian and Aboriginal history.

LIMITED EDITION OF 100 COPIES, 124 pages, illustrated with Idriess timeline, numbered, in colour; for more inform-ation write to ettimprint@hotmail.com

Paperback edition, black and white photographs throughout, 124 pages, illustrated. Out now.

www.ingramcontent.com/pod-product-compliance
Lightning Source LLC
Chambersburg PA
CBHW031942190326
41519CB00007B/626